IEEE Radio and Antenna Days of the Indian Ocean (IEEE RADIO 2017)

Extended Papers

IOP Conference Series: Materials Science and Engineering
Volume 321

Cape Town, South Africa
25 - 28 September 2017

Editor:

Vikass Monebhurrun

ISBN: 978-1-5108-6038-4
ISSN: 1757-8981

Printed from e-media with permission by:

Curran Associates, Inc.
57 Morehouse Lane
Red Hook, NY 12571

Some format issues inherent in the e-media version may also appear in this print version.

Printed by Curran Associates, Inc. (2018)

For permission requests, please contact the Institute of Physics at the address below.

Institute of Physics
Dirac House, Temple Back
Bristol BS1 6BE UK

Phone: 44 1 17 929 7481
Fax: 44 1 17 920 0979

techtracking@iop.org

Additional copies of this publication are available from:

Curran Associates, Inc.
57 Morehouse Lane
Red Hook, NY 12571 USA
Phone: 845-758-0400
Fax: 845-758-2633
Email: curran@proceedings.com
Web: www.proceedings.com

Table of contents

Volume 321

IEEE Radio and Antenna Days of the Indian Ocean (IEEE RADIO 2017)

25–28 September 2017, South Africa

Accepted papers received: 19 February 2018
Published online: 22 March 2018

Preface

Papers

IOP Conf. Series: Materials Science and Engineering **321** (2018) 011001 doi:10.1088/1757-899X/321/1/011001

EDITOR'S FOREWORD

Professor Weng Cho Chew, IEEE Antennas and Propagation Society (APS) President-Elect 2018, officially opened the fifth edition of the 2017 IEEE Radio and Antenna Days of the Indian Ocean (IEEE RADIO 2017) international conference that was held at the Protea Hotel by Marriott Cape Sea Point, Cape Town, South Africa from 25th to 28th September 2017. The conference was organized by the Radio Society (Mauritius) and financially sponsored by IEEE APS. IEEE RADIO 2017 was also technically co-sponsored by Union Radio Scientifique Internationale (URSI) and IEEE Region 8.

The local organizing committee comprised researchers from CentraleSupélec, France, Université de La Réunion, University of Mauritius and Square Kilometre Array (SKA) Africa. We are grateful to SKA Africa, Newton Fund, Development in Africa for Radio Astronomy (DARA), Mobile & Wireless Forum (MWF) and Computer Simulation Technology (CST) for providing financial sponsorship for the conference.

Participants from over twenty different countries spanning all the five continents, attended IEEE RADIO 2017. The conference featured two workshops and twelve regular sessions. The conference covered diverse topics such as radio astronomy, antenna design, wave propagation and scattering, couplers, waveguides and resonators, wireless communication systems, devices and circuits, computational electromagnetics, and biological effects and medical applications of electromagnetic fields. The first workshop, held on the first day of the conference, was dedicated to the global SKA project which extends to the Indian Ocean islands. On the last day of the conference, the second workshop addressed the technical and financial implications of counterfeit mobile devices. Invited talks were delivered by several internationally recognized experts and scientists.

The oral presentations of young scientists and students were evaluated by a jury. A cash prize of 300 Euros and a plaque were awarded to the Best Young Scientist less than 36 years old. Moreover, three cash prizes of 300 Euros, 200 Euros and 100 Euros as well as plaques were awarded to the first, second and third Best Student papers, respectively.

All the 2-page long papers that were effectively presented during the conference were submitted to the IEEE Xplore Digital Library. As usual, the authors of the accepted and presented conference papers were invited to submit an extended version (limited to a maximum of 4 pages) for publication as a volume in the IoP Conference Series: Materials and Engineering (MSE). The eighteen papers that were submitted were peer-reviewed by the distinguished members of the Scientific Committee. Only ten papers were effectively accepted. A major cause of rejection was insufficient additional material compared to the 2-pages paper already published on the IEEE Xplore Digital Library

Publication of the extended papers for the fifth time in the IoP Conference Series: MSE bears testimony to the high technical quality that is targeted by the RADIO series of international conferences.

Editor
Vikass Monebhurrun

IOP Conf. Series: Materials Science and Engineering **321** (2018) 011002 doi:10.1088/1757-899X/321/1/011002

IOP Conf. Series: Materials Science and Engineering **321** (2018) 011003 doi:10.1088/1757-899X/321/1/011003

Peer review statement

All papers published in this volume of *IOP Conference Series: Materials Science and Engineering* have been peer reviewed through processes administered by the proceedings Editors. Reviews were conducted by expert referees to the professional and scientific standards expected of a proceedings journal published by IOP Publishing.

IOP Conf. Series: Materials Science and Engineering **321** (2018) 012001 doi:10.1088/1757-899X/321/1/012001

Phased-locked signal detection system based on parallel optimal receivers by average risk criterion

D A Dolmatov, R R Abdullin and R I Sokolov

Engineering School of Information Technologies, Telecommunications and Control Systems, Ural Federal University named after the first President of Russia B.N.Yeltsin, 19 Mira Street, Yekaterinburg, Russian Federation

r.r.abdullin@urfu.ru

Abstract. This paper is devoted to the investigation of a developed sync system which is based on optimal receiver by average risk minimum criterion when detecting the reflected radar impulse under conditions of Gaussian and Johnson noises. M-sequence with single pulse phase-shift keying has been chosen as a reference signal. A transceiver has been implemented during the digital experiment. The use of developed self-tuning system allows applying the optimal methods of reflected radar signal detection when it is noncoherently received under conditions of Gaussian and non-Gaussian noises. It has been proved that the developed sync system has the signal detection quality advantage over the system based on matched filter.

1. Introduction

The development of radar signal detection systems, which allow defining the object at the long distance, is still a problem today [1]. Existing detection systems are generally based on the methods of correlation analysis and matched filtering with preliminary adaptive or nonlinear filtering [2]. The use of optimal reception methods requires the huge amount of *a priori* data on signal and noise. For example, the use of phase-shift keying (PSK) implies that receiver knows the single pulse duration, signal amplitudes, as well as the arrival time and initial phase of the first pulse [3]. This is impossible because of *a priori* uncertainty about a target range (the time of propagation in space) and the signal distortions under the conditions of different interferences. Therefore, the necessity arises to develop the sync system locking the first pulse initial phase.

For this purpose in the present work we suggest using maximal value of cross-correlation function (CCF) between the reference signal and the signal from the optimal receiver output. We considered a discrete random process with Johnson distribution as the noise since a variation of its parameters allows generating a sufficiently large class of distributions close to the distribution of real noise [4]. Moreover, the choice of Johnson distribution is convenient because these processes can be generated by inertialess nonlinear transformations of normal random processes, in particular discrete.

2. Synthesis of Optimal Algorithm

The task of forming the optimal reception algorithm for reflected radar signal in the sum with non-Gaussian noise is solved based on the following key assumptions [5]:

1. The signal $y(t)$ is M-sequence with PSK, whose parameters (amplitude, duration, repetition period, the carrier frequency) are known except the message $\lambda(t)$.

2. The noise is a discrete sequence coinciding in time with the signal samples and having the Gaussian or Johnson distribution. The noise intensity is so high, that the signal to noise ratio (SNR) is much less than unity. Thus, the signal accumulation is used for reliable information extraction.

3. The criterion of the average risk minimum is used as an optimality criterion for a simple loss function. Therefore, the algorithm of *a posteriori* probability density maximum in the form of likelihood ratio can be used as an optimum decision rule. In this criterion the optimal reception provides the minimum of total error probability.

During the present work we considered the optimal signal receiver under the conditions of additive non-Gaussian noise when SNR is quite small. In this case the noise is a random process described as an inertialess transform $q(z_i)$ of Markov Gaussian process z_i with variance σ^2. The input process $y(t)$ is assumed to be observed at discrete moments of time t_i:

$$y_i = \begin{cases} s_i + q(z_i), \mathit{if}\ \Theta = 1, \\ q(z_i), \mathit{if}\ \Theta = 0, \qquad i = 1, 2, ..., m. \end{cases}$$

Parameter Θ characterizes the presence ($\Theta = 1$) or absence ($\Theta = 0$) of the signal s_i in observed data. In turn the useful signal s_i is a harmonic signal with two states of phase differed by π depending on parameter Ω, which means the positive ($\Omega = 1$) or negative ($\Omega = 0$) pulse of M-sequence.

It is supposed the function q is known, and it has such unique inverse function $q^{-1} = Q$, that

$$z_i = \begin{cases} Q(y_i - a_i \sin(\omega t)), \mathit{if}\ \Theta = 1, \mathit{and}\ \Omega = 1, \\ Q(y_i - a_i \sin(\omega t + \pi)), \mathit{if}\ \Theta = 1, \mathit{and}\ \Omega = 0, \\ Q(y_i), \mathit{if}\ \Theta = 0. \end{cases}$$

where a_i is an amplitude of every single pulse; ω is a cyclic frequency of radio pulse filling. Then initial $w_0(z)$ and conditional $w(z_{i+1}|z_i)$ probability densities are defined as follows [6]:

$$w_0(z) = \frac{1}{\sigma_z \sqrt{2\pi}} \exp\left\{ -\frac{Q^2(y_0)}{2\sigma_z^2} \right\} |Q'(y_0)|,$$

$$w(z_{i+1}|z_i) = \frac{|Q'(y_{i+1})|}{\sigma_z 2\pi \sqrt{1-R^2}} \exp\left\{ -\frac{\left[Q(y_{i+1}) - Q(y_i) R \right]^2}{2\sigma_z^2 (1-R^2)} \right\},$$

where R is a correlation coefficient. The optimal receiver is built by modelling the likelihood ratio [7]. Since the signal contains two types of useful message, it is necessary to implement two optimal algorithms for $\Omega = 1$ and $\Omega = 0$. So the overall likelihood equation looks as follows

$$\Lambda_m = \frac{w_0(z_0, \Theta = 1) \prod_{i=1}^{m} w(z_{i+1}, \Theta = 1 | z_i, \Theta = 1)}{w_0(z_0, \Theta = 0) \prod_{i=1}^{m} w(z_{i+1}, \Theta = 0 | z_i, \Theta = 0)}. \qquad (1)$$

Equation (1) fully defines the receiver scheme presented in figure 1. The signals from the adder output go to the total adder and then to the solver whose answer is "signal_1" or "signal_2" depending on whether incoming message crosses threshold l_{m1} or l_{m2} respectively, and 0 otherwise.

3. Synchronization Algorithm Synthesis

An implementation of the optimal receiving algorithm is only possible with the precise *a priori* data on the first pulse arrival. Therefore, it is proposed to use the parallel processing of the received signal and its copies shifted in time by the sample from each other in the range of half number of samples per pulse. From the simple processor output the restored sequence goes to the correlator calculating the maximal value of CCF between the reference and restored signals (figure 2). Then, the correlator and appropriate processing channel with the largest possible CCF are defined. The clock shift at the chosen channel guarantees the best quality of restoration and synchronization.

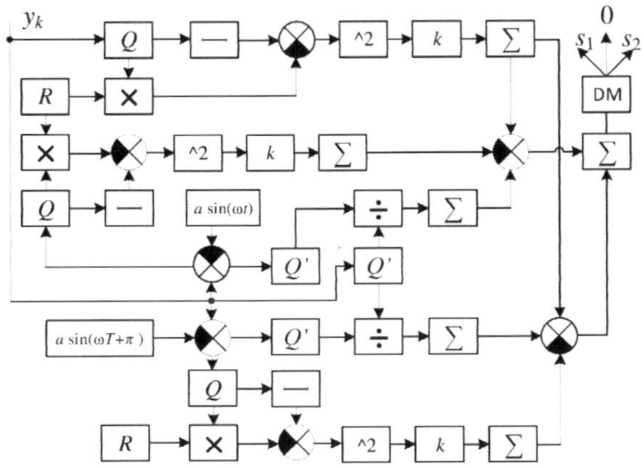

Figure 1. The optimal receiver block diagram.

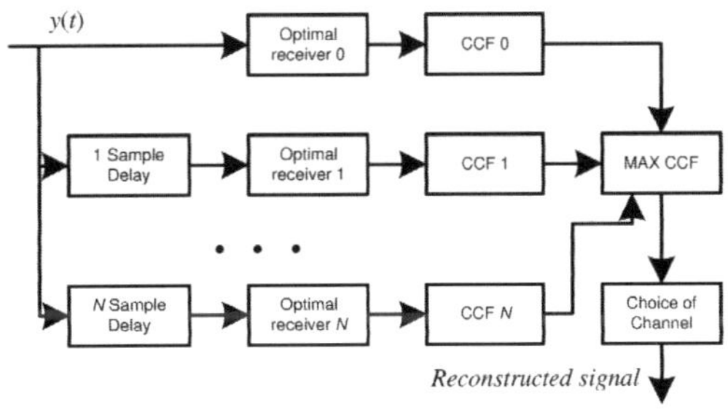

Figure 2. Block diagram of sync system.

4. Digital Implementation of Sync System

The developed algorithms of optimal reception and sync system have been implemented in LabView software set up on the National Instruments platform PXI 1075. Figure 3 shows the maximal values of different CCFs when parallel processing the reference signal and 12 signal copies shifted in time by sample from the previous one. CCF is maximal when the initial phases of the reference and received signals coincide. The optimal processing algorithm is launched at this moment and performs the best.

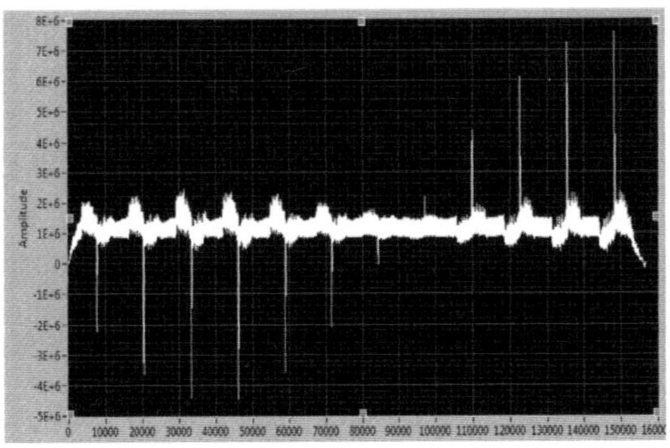

Figure 3. The determination of the largest possible CCF between the reference signal and received signal with the time shift.

Figure 4 shows the signal and noise mixture at the receiver input, and the signal from the phase-locked output. In order to estimate the restoration quality we have obtained the dependencies of the normalized CCF (NCCF) vs. SNR for the developed system and compared them with matched filter under white Gaussian noise (WGN) and different types (S_B, S_L and S_U) of Johnson noise (figure 5).

5. Conclusion

The optimal restoration algorithm for radar signal, represented as M-sequence with PSK, has been developed for operation under conditions of Gaussian and non-Gaussian noise. The implementation of suggested algorithm requires the use of precious *a priori* data on received signal initial phase. Thus, the self-tuning system, which is based on the parallel processing of several signal copies shifted in time, has also been designed.

The use of optimal algorithm leads to the gain in signal restoration by 10 dB for WGN and 2 dB for S_L and S_U Johnson noise in comparison with the use of matched filters. The worst masking effect belongs to S_B Johnson noise. At the same time the use of optimal algorithm in this case does not lead to the significant gain in signal restoration.

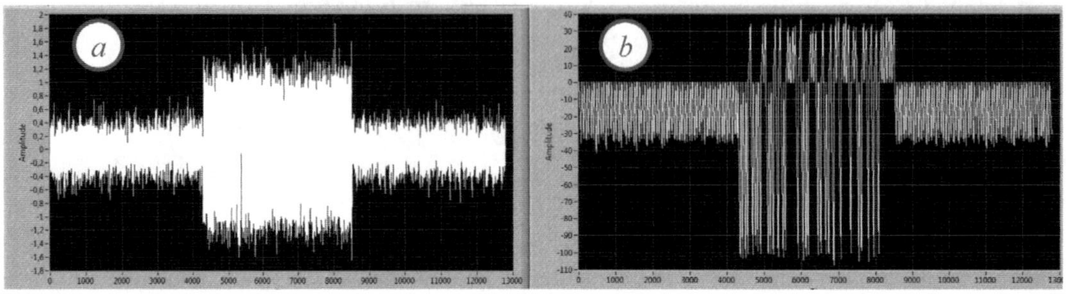

Figure 4. The result of developed system operation: (*a*) mixture of signal and WGN mixture at the input; (*b*) restored signal. SNR = 5 dB.

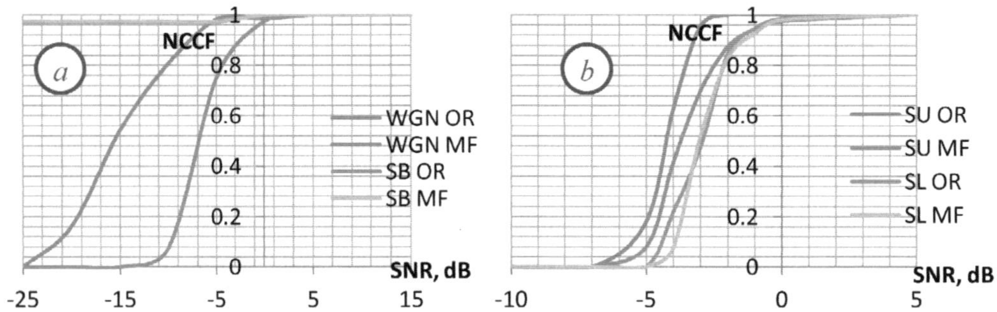

Figure 5. The dependencies of NCCF vs. SNR at the output of matched filter (MF) and optimal receiver (OR) for (*a*) WGN and S_B Johnson noise; (*b*) S_L and S_U Johnson noise.

Acknowledgments

The work was supported by Act 211 Government of the Russian Federation, contract № 02.A03.21.0006.

The research was executed by the grant of the Ministry of education and science of the Russian Federation (project № 8.2538.2017/4.6).

References

[1] Bakulev P A 2004 *Radar systems: Textbook for Universities*. (Moscow: Radiotechnics)

[2] Sokolov R I and Abdullin R R 2015 Synthesis of ultra-wideband signals receiver algorithm based on Markov theory of nonlinear filtering *Proc 2015 IEEE International Conference on Microwaves, Communications, Antennas and Electronic Systems (Tel-Aviv)*

[3] Dmitriev V D 2003 Technology transfer information using ultra-wideband signals (UWB) *Components and Technologies*, no. 9 – http://www.kit-e.ru/assets/files/pdf/2003_09_72.pdf.

[4] Kendall M G and Stuart A 1977 *The advanced theory of statistics: distribution theory.* (Macmillan)

[5] Gutkin L S 1972 *The theory of optimal receiving methods with fluctuation interference.* (Moscow: Sov. Radio)

[6] Sokolov R I and Abdullin R R 2016 Research of optimal pulse signal reception quality by mean risk minimum criterion with action of Gaussian and non-Gaussian noise *Proc 2016 International Conference on Computational Techniques in Information and Communication Technologies (New Delhi)*

[7] Astretsov D V, Nifontov Yu A and Sokolov R I 2012 Analyzes of potencial noise immunity binary signal extraction in condition of Gaussian and non-Gaussian noise *Proc. Physics and technical applications of wave processes* (Ekaterinburg: UrFU) pp 147–149

IOP Conf. Series: Materials Science and Engineering **321** (2018) 012002 doi:10.1088/1757-899X/321/1/012002

Slotted rectangular waveguide with dielectric sandwich structure inside

R R Abdullin and R I Sokolov

Engineering School of Information Technologies, Telecommunications and Control Systems, Ural Federal University named after the first President of Russia B.N.Yeltsin, 19 Mira Street, Yekaterinburg, Russian Federation

r.r.abdullin@urfu.ru

Abstract. This paper continues the series of works devoted to the investigation of leaky-wave antenna based on layered rectangular waveguide with periodic transverse slots in broad face. Previously developed wavenumber calculation technique has been adapted for analysis of slotted sandwich waveguide with three layers at least. The paper provides the numerical results of velocity factor dependencies for partially filled slotted rectangular waveguide containing a dielectric slab in the middle position inside or an air gap between two dielectric slabs. Additionally, dispersion properties are also considered for multilayer waveguide with linear laws combinations of thickness and permittivity. This allows recognizing the trends to develop new prospective antennas with complex patterns of tilt angle change. All numerical results obtained are confirmed with the in-situ measurements of transmission coefficient phase.

1. Introduction

Recently the new model of leaky-wave antenna based on rectangular waveguide with transverse slots in broad face and sandwich structure inside has been suggested in [1]. The technique of composing the dispersion equation has been developed. The numerical results of its solution have been obtained in [2]: frequency dependencies of velocity factor, radiation pattern tilt angle, and beamwidth. The experiments described in [3] have confirmed the validity of the calculation and mathematical model.

However, these studies were limited only by consideration of two-layered structure and analysis of waveguide filling factor when dielectric was immediately at the face parallel to the slot plane. This fact does not show a complete picture of the process inside the partially filled slotted waveguide.

The use of three layers in the model can be helpful for analysis of the dielectric slab position impact on the antenna performance. Using the bigger number of layers allows establishing required patterns of change in permittivity, permeability, and thickness between adjacent slabs to solve the various tasks. For instance, it can provide the linear law of angle-frequency performance variation, which is important for possibility of these antennas application in airborne radars.

2. Investigated Antenna Model

The model of investigated antenna is presented in figure 1. This is the endless slotted rectangular waveguide with cross section dimensions $a \times b$. The slotted face is combined with an infinite top ground plane. Transverse slots are located at the distance of x_0 from narrow face and p from one another. Length and width of slots are L and W, respectively. The waveguide contains the sandwich structure inside, whose layers are parallel to the ground plane. Every slab has its own thickness d_p, permittivity ε_p, and permeability μ_p. For the convenience wall thickness is taken as infinite small and wall material is a perfect electric conductor. Coordinate axes are chosen as follows: y is along the waveguide axis, z is normal to the slot plane, and x is oriented to complete the right-hand system.

This system is supposed to support the propagation with wavenumber $k_y = \beta - j\alpha$, where β is a phase constant and α is an attenuation constant. The complex propagation constant k_y is defined from a transcendental equation by matching the tangential components of magnetic field from different sides of the slot plane according to the technique offered in [1] and [4].

The investigated antenna is divided into two subspaces (external and internal), either of which is described by equivalent lines model. This model uses the Green's function method and allows taking into account the structure inhomogeneity. Generally, dispersion equation for leaky-wave antenna based on multilayer waveguide looks like equation for antenna on the basis of two-layered waveguide:

$$\sum_{n=-\infty}^{\infty}\int_0^{\infty}\frac{1}{k_x^2+k_{yn}^2}\left[k_x^2\vec{Y}_{\text{ext}}^H(0)+k_{yn}^2\vec{Y}_{\text{ext}}^E(0)\right]\left[\frac{\cos(k_xL/2)}{(\pi/L)^2-k_x^2}\right]^2 dk_x = \tag{1}$$

$$=\frac{2\pi}{a}\sum_{n=-\infty}^{\infty}\sum_{m=0}^{\infty}\frac{1}{(\pi m/a)^2+k_{yn}^2}\left[\frac{\varepsilon_m}{2}(\pi m/a)^2\vec{Y}_{\text{int}}^H(0)+k_{yn}^2\vec{Y}_{\text{int}}^E(0)\right]\cdot\left[\frac{\sin(\pi m x_0/a)\cos(\pi mL/2a)}{(\pi m/a)^2-(\pi/L)^2}\right]^2,$$

where k_x is wavenumber along the axis x in space region; $k_{yn} = k_y + 2\pi n/p$, where n is an order of space Floquet harmonic; ε_m is Kronecker delta taking a value $\varepsilon_m = 1$, if $m = 0$, and $\varepsilon_m = 2$, if $m \neq 0$; $\vec{Y}_{\text{ext,int}}^H(0)$, $\vec{Y}_{\text{ext,int}}^E(0)$ are external and internal input modal admittances from the slot plane in H-line and E-line, respectively. However, the one substantial difference takes a place: the bigger the number of layers, the bigger iterations quantity is required for calculation of input admittances (figure 2) [5]:

$$\vec{Y}^{E,H}(z_p)=Y_{p+1}^{E,H}\frac{\vec{Y}^{E,H}(z_{p+1})\cot\gamma_{p+1}d_{p+1}+jY_{p+1}^{E,H}}{Y_{p+1}^{E,H}\cot\gamma_{p+1}d_{p+1}+j\vec{Y}^{E,H}(z_{p+1})}, \tag{2}$$

where $Y_p^H = \gamma_p/\omega\mu_p$ and $Y_p^E = \omega\varepsilon_p/\gamma_p$ are equivalent wave admittances of internal layers; $\gamma_p = \sqrt{k_p^2-(\pi m/a)^2-k_{yn}^2}$ is a longitudinal propagation constant of every single dielectric slab in chosen coordinate system, m is a number of half waves along the broad side of waveguide.

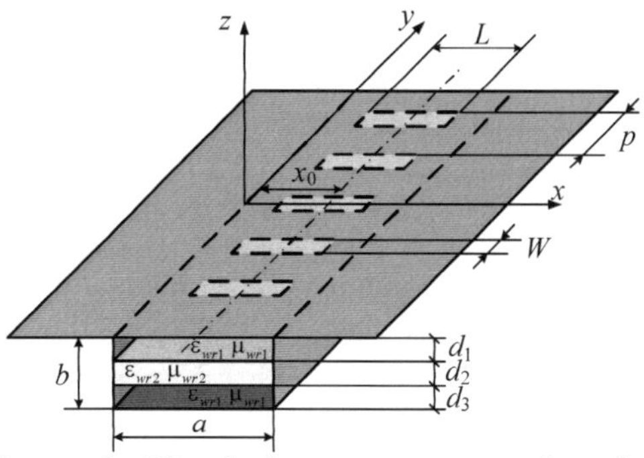

Figure 1. The leaky-wave antenna based on rectangular waveguide with sandwich structure

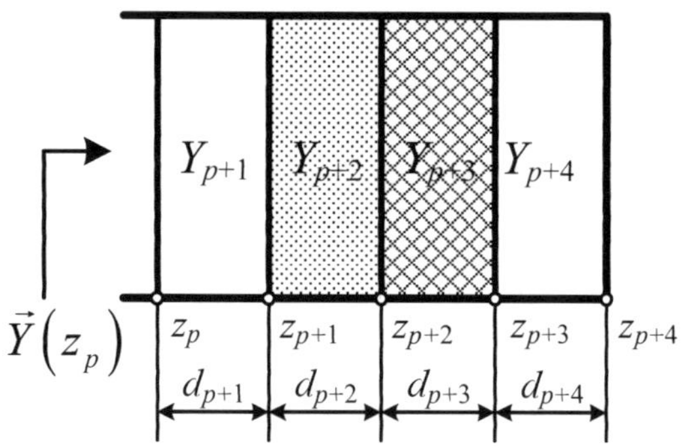

Figure 2. For the admittances recalculation.

3. Numerical Results

The equation (1) is transcendental since it contains nonalgebraic functions, particularly trigonometric. That is why its solution is only possible by means of numerical methods. The standard parameters we used are following: $a = 23$ mm, $b = 10$ mm, $L = 8$ mm, $W = 0.5$ mm, $p = 2$ mm, $x_0 = a/2 = 11.5$ mm.

IOP Conf. Series: Materials Science and Engineering **321** (2018) 012002 doi:10.1088/1757-899X/321/1/012002

3.1. Partially Filled Rectangular Waveguide

In previous works [2] and [3] it has been shown the structure has bigger value of velocity factor if the dielectric slab is located near the slotted wall in comparison with position near the opposite one. However, phase velocity dependencies for waveguide with dielectric in the middle position were unknown. Figure 3 shows the plot of velocity factor for leaky-wave antenna based on rectangular waveguide partially filled with Teflon ($\varepsilon = 2$) at frequency of 8 GHz. Figure 3*a* illustrates the dependence for dielectric slab shifted from the face opposite to slots, whereas figure 3*b* demonstrates the results for inverted structure, i.e. for an air gap between two dielectric slabs.

Really, the closer dielectric to slots, the smaller value of phase velocity. In both cases curves are seen to have nonlinear patterns of change with dielectric (or gap) shift directional extremums. And the more the dielectric layer thickness, the more dependencies distortion.

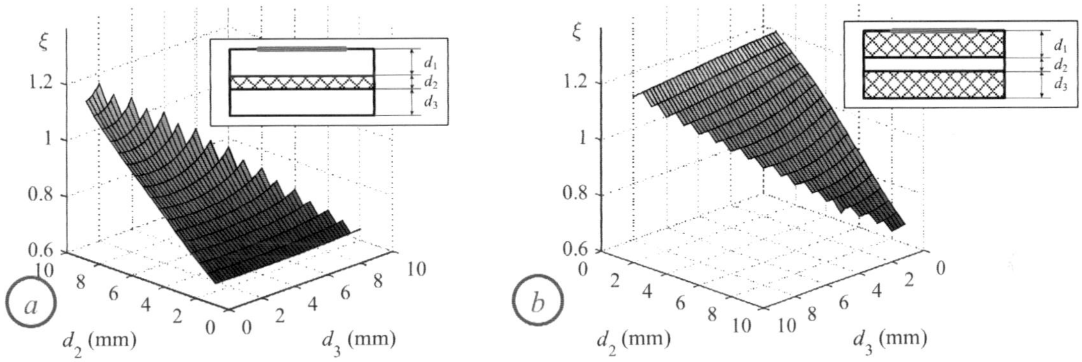

Figure 3. Two-dimensional velocity factor dependence of slotted waveguide on thickness and position of (*a*) dielectric slab; (*b*) gap between two dielectric slabs.

3.2. Multilayer Slotted Rectangular Waveguide

The following part of paper is devoted to the investigation and discussion of different laws of change in thickness and permittivity between adjacent layers inside the waveguide, and definition of their influence on the antenna performance. At the first stage of research only simple linear laws have been considered. Figure 4 shows dispersion dependencies for leaky-wave antenna based on rectangular waveguide filled with five dielectrics whose parameters follow in different orders. Permittivity changes from 1 to 5, more dark shading corresponds to bigger values of permittivity. Figure 5 shows analogous situation with the change in dielectric layer thickness.

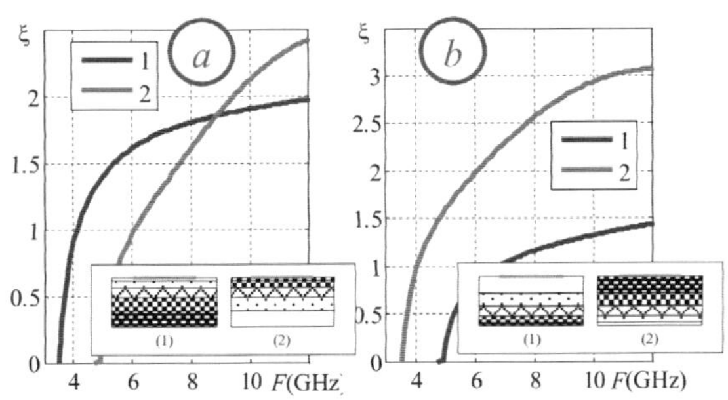

Figure 4. Dispersion curves for multilayer waveguide with periodic transverse slots at the linear increase (*a*) and decrease (*b*) of dielectric layer thickness.

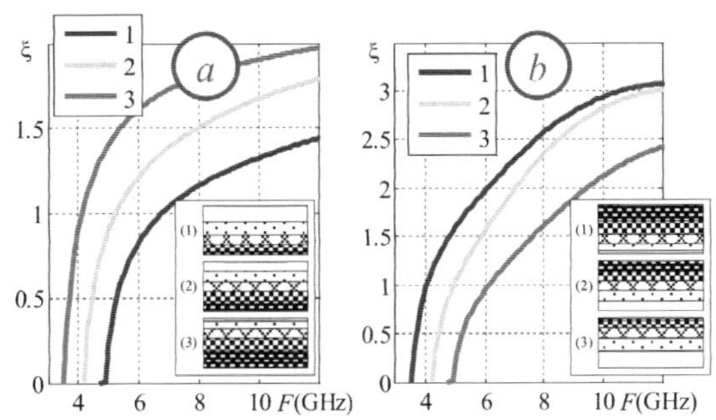

Figure 5. Dispersion curves for multilayer waveguide with periodic transverse slots at the linear increase (*a*) and decrease (*b*) of dielectric layer permittivity.

The results obtained for multilayer waveguide confirm and complete the trends obtained for partially filled waveguides. The curves from figures 4 and 5 look quite similar and make it possible to conclude following. An arithmetic mean value of permittivity inside the waveguide lies behind the system cut-off frequency. At the same time the bigger the permittivity of layer located near the slots, the higher the slope of dispersion curve at a point of cut-off frequency.

4. Experimental Results

In order to prove the correctness of developed calculation technique and obtained numerical results we have conducted experimental research, which implies some in-situ measurements of transmission coefficient S_{21}. We used the same workpiece of slotted waveguide and the same dielectric stripes from Teflon and PCT-5 as it had been in [3]. We placed it in different order inside the waveguide and built the velocity factor dependencies recalculated from a phase of transmission coefficient S_{21}. Figure 6 shows the studied cases of cross section and corresponding experimental dispersion curves in comparison with the theoretical data. It is seen they have a good agreement.

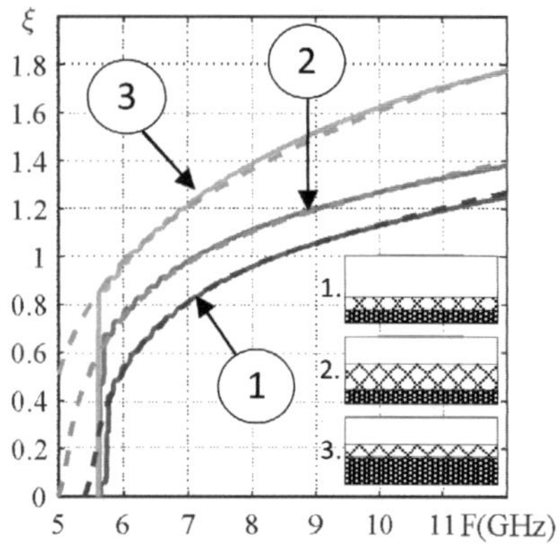

Figure 6. Comparison of experimental and theoretical data. Solid lines – measured values; dashed lines – calculated dependencies. More dark shading in sketches corresponds to PCT-5 (ε = 5), light shading is to Teflon (ε = 2).

Some inaccuracy takes a place in a low part of studied frequency band because of very high slope of dispersion curve such the phase changes so quickly, whereas a frequency step is too low. The other possible reason is a feature of calibration with coaxial-to-waveguide transition.

5. Conclusion

The velocity factor calculation technique has been adapted for analysis of leaky-wave antenna based on rectangular waveguide with periodic transverse slots and internal sandwich structure. The dielectric layer position influence on the leaky-wave antenna propagation constant is considered in case of partially filled waveguide. The dielectric slab thickness significantly impacts on the dispersion curves distortion. Also, antenna based on five-layered structure has been investigated. All theoretical results are confirmed with experimental research. The combination of dielectric layers with various parameters allows obtaining arbitrary performance of periodic leaky-wave antenna.

Acknowledgements

The work was supported by Act 211 Government of the Russian Federation, contract № 02.A03.21.0006.

References

[1] Abdullin R R, Knyazev S T, Lesnaya L L and Shabunin S N 2013 Analysis of partially dielectric-filled rectangular waveguide with transverse slots using Green's function method *Proc 7th European Conference on Antennas and Propagation (Gothenburg)* pp 3453–57.

[2] Abdullin R R, Mitelman Yu E and Shabunin S N 2014 Radiation pattern of leaky-wave antenna based on partially-filled rectangular waveguide. *Proc Loughborough Antenna & Propagation Conference (Loughborough)* pp 516–518.

[3] Abdullin R R 2016 Experimental research of leaky-wave antenna based on partially-filled rectangular waveguide with transverse slots *Anttennas* no 3 (Moscow: Radiotechnics) pp 46-54.

[4] Liu J, Jackson D R and Long Y 2011 Modal analysis of dielectric-filled rectangular waveguide with transverse slots *IEEE Trans. Antennas and Propag.*, vol. 59, pp 3194-3203.

[5] Panchenko B A, Nefedov E I 1986 *Microstrip antennas*. (Moscow: Radio & Connection).

IOP Conf. Series: Materials Science and Engineering **321** (2018) 012003 doi:10.1088/1757-899X/321/1/012003

Electromagnetic analysis and preliminary commissioning results of the shaped dual-reflector 32-m Ghana radio telescope

M Venter[1] and P Bolli[2]

[1] SKA South Africa, Pinelands, South Africa, 7405
[2] INAF - Arcetri Astrophysical Observatory, Firenze, Italia, 50125

E-mail: mventer@ska.ac.za

Abstract. This paper presents results from the electromagnetic analysis of the African VLBI Network shaped Ghana radio telescope at the operating frequencies of 5 and 6.7 GHz. The geometry implemented in commercial electromagnetic software provides insight into the effects of the slanted beam-waveguide, shaped reflector illumination and mechanical tolerances, which are known to be more stringent compared to a perfect paraboloid. It is shown that the theoretical maximum gain and aperture efficiency at 5 GHz are 63.80 dBi and 85.45%, respectively. The corresponding values at 6.7 GHz are 66.47 dBi and 88.00%, respectively. Comparisons to sidelobe maps produced from astronomical observations are also discussed, showing possible misalignment in the structure when utilised outside its originally intended purpose.

1. Introduction

The African Very Long Baseline Interferometry (VLBI) Network (AVN) is a Square Kilometre Array South Africa (SKA SA) initiative funded by the African Renaissance Fund (ARF). This project aims to create science opportunities in participating countries and enable partaking in SKA pathfinder development by amongst others, converting large antennas into radio astronomy telescopes. The early electromagnetic analysis of a 32-m telecommunications dish in Kutunse, Ghana, the first conversion effort of the project, is described in [1]. The effort is expanded upon here, with analysis and characterisation of the effects of the shaped profiles at 5 and 6.7 GHz. Additionally, the simulated antenna patterns are compared with results from the 'first light' science observations.

2. Overview of antenna optics

The Ghana antenna consists of a Cassegrain shaped dual-reflector system with four mirrors in a 12.68° slanted beam-waveguide (BWG). A full discription of the antenna optics is provided in [1] with the theory on shaping synthesis of dual-reflector antennas detailed in [2]. The primary reflector has an f/d of 0.32 and the Cassegrain system an equivalent focal length of 107.68 m. The shaped sub-reflector has a foci length and half-subtended angle of approximately 9.44 m and 8.50°, respectively. The maximum difference between the best-fit parabolic profile and the shaped primary mirror and the best-fit hyperbolic profile and the shaped sub-reflector is 35 mm and 155 mm, respectively.

IOP Conf. Series: Materials Science and Engineering **321** (2018) 012003 doi:10.1088/1757-899X/321/1/012003

3. Performance results

This section introduces results obtained by solving the electromagnetic problem with the physical optics method in GRASP (General Reflector Antenna Software Package) [3]. The system is excited with a Gaussian feedhorn with a -22 dB taper, as specified in antenna documentation [4], where the large taper minimizes the spillover in the BWG. Fig. 1 shows the 6.7 GHz co-polar radiation patterns for the E-plane and H-plane of the shaped antenna, where Table 1 shows comparison between the performance of the shaped and equivalent unshaped case. The cross-polarisation, sidelobe level (SLL) and beam near horizon values are given relative to the maximum gain at zenith. The shaped design has the effect of narrowing the mainbeam for the given reflector size(s), which can be seen in a gain increase of approximately 0.7 dB and reduction in total spillover. However, the spillover in the BWG is a major contributor of an increase in system noise as the shroud surrounding the mirrors are seen by the antenna at ground temperature (here assumed as 290 K). Consequently, it was found that the BWG would add approximately 5.5 K noise at 6.7 GHz, which in radio astronomy terms is not insignificant. Also, a disadvantage in the shaped design when used for radio astronomy is the significant increase in SLL. Furthermore, because of the asymmetry between the curved BWG mirrors, the far field rotates by 25.3° in azimuth (twice the degree of the slant rotation) which results in less prominent nulls near the mainbeam as well as significant depolarisation which is the major impact of the BWG.

Table 1. Ghana antenna performance values comparison between shaped and unshaped system.

Configuration	Unshaped		Shaped	
Parameter	5 GHz	6.7 GHz	5 GHz	6.7 GHz
Maximum Gain (dBi)	63.09	65.80	63.80	66.47
Maximum Cross-polarisation (dB)	−36.98	−37.92	−36.74	−37.77
Half-power Beamwidth (HPBW)(°)	0.13	0.10	0.11	0.09
SLL (dB)	−22.63	−22.90	−15.21	−15.15
Beam Near Horizon (dB)	−59.11	−53.52	−48.09	−53.68
Spillover (dB)	0.55	0.45	0.19	0.14
Aperture Efficiency (%)	72.56	75.42	85.45	88.00

4. Antenna efficiency

The shaped profiles performs its purpose in that it delivers a high gain, high aperture efficient reflector antenna. This is accomplished by the inverse taper of the illumination distribution on the main reflector as shown in Fig. 2, where higher power towards the unblocked part of the reflector is visible. The asymmetrical illumination caused by the slant in the BWG is also noticeable. The Ghana antenna has an illumination efficiency of approximately 92% at 6.7 GHz, which is the product of the taper efficiency (94%) and the phase efficiency (97%). The aperture field phase (represented as a surface deviation plot and assuming no surface roughness) is shown in Fig. 3 and is not particularly flat in the region where the illumination map shows its highest power. It seems the middle area of the dish is out by 0.5 mm compared to the outer part which has higher amplitude values. It is clear that the shaping utilises less the center part of the dish as one would expect to see with a conventional parabolic profile.

5. Mechanical tolerances

The Ghana antenna was originally optimised for a single elevation of approximately 64° when observing a geostationary satellite, which raises uncertainties when utilising the structure to

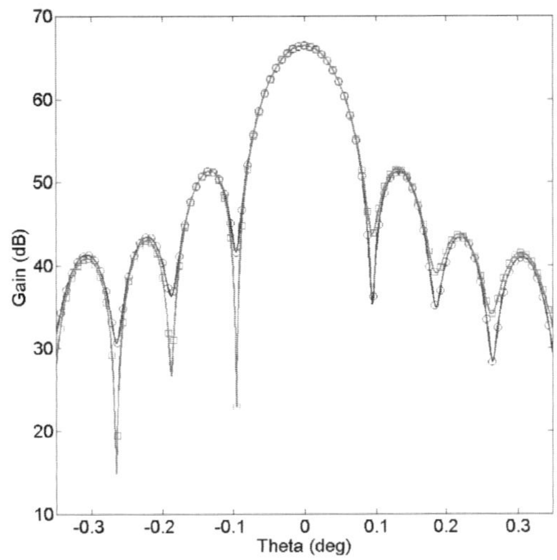

Figure 1. E-plane (blue circles) and H-plane (red squares) patterns of the Ghana shaped antenna.

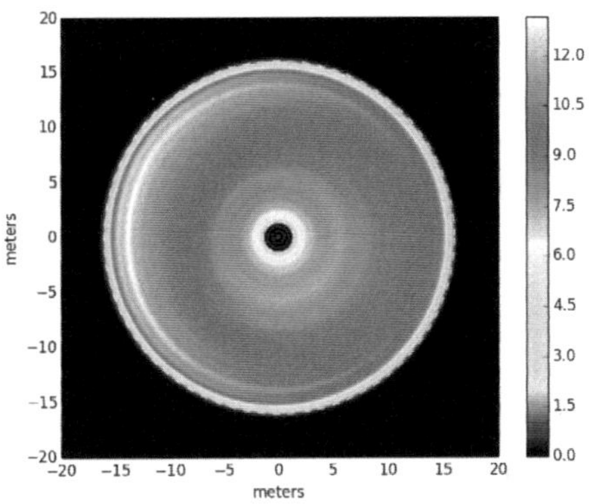

Figure 2. Aperture field distribution plot of the Ghana shaped antenna at 6.7 GHz.

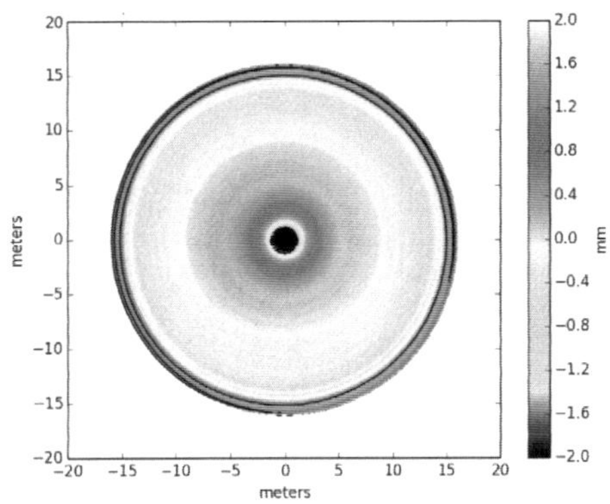

Figure 3. Surface deviation plot of the Ghana shaped antenna at 6.7 GHz.

observe at many different elevations. Especially, displacement of the sub-reflector in the primary focus position will have the greatest effect on the performance of the antenna. GRASP simulations confirms this with displacement of the sub-reflector attributing 94% of gain loss when considering the average contribution of all six mirrors for four different displacement values.

6. Comparison to commissioning results

An useful step towards realising the actual alignment condition of the antenna, is to perform observations of several strong astronomical sources. This is done in an attempt to develop sidelobe maps of the mainbeam at the higher operating frequency of 6.7 GHz, which can be used as a diagnostic for possible misalignment of the sub-reflector. An example of the results are shown in Fig. 4 where observations of Cygnus A (a radio galaxy) and Taurus A (Crab Nebula supernova remnant) are compared to simulations of on-axis displacement of the sub-reflector. Taurus A is the stronger but more extended source, resulting in a broadening of the mainbeam and filled in nulls. Cygnus A in comparison, is weaker but more compact. These observations indicate that the sidelobes are reasonably symmetrical but higher than indicated by the simulations, which can likely be attributed to a focusing error.

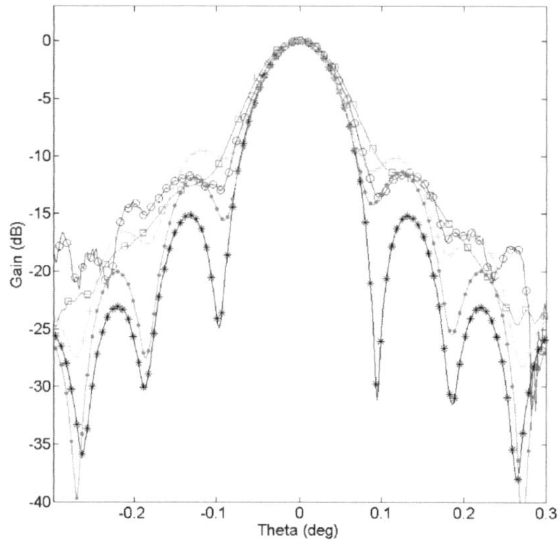

Figure 4. Beam pattern simulations for the Ghana dish in comparison with Cygnus A (blue circles) and Taurus A (red squares) observations. The GRASP patterns shown are those without any sub-reflector offsets (black asterisks) and offsets of plus (green pluses) and minus (magenta dots) 2.25 cm in the up-and-down zenith direction.

7. Conclusion

This paper gives a compact overview of the performance of a shaped dual-reflector antenna and its effects in performance when utilised for radio astronomy purposes. As very little is known about the real aperture efficiency of the Ghana antenna, future work includes physical measurement of the primary and sub-reflector to determine their roughness using the microwave holography technique. This will also determine misalignments in the structure which can be corrected to ensure optimal performance.

Acknowledgments

The authors acknowledge Christophe Granet from Lyrebird Antenna Research for providing an approximate shaping geometry. The authors also acknowledge George Nicolson from HartRAO, South Africa, the AVN commissioning team and Mattieu de Villiers from SKA SA. With support from the National Research Foundation (NRF) and the Department of Science and Technology (DST) of South Africa and Italy's Ministry of Foreign Affairs and International Cooperation, Directorate General for the Country Promotion (Bilateral Grant Agreement ZA14GR02 - Mapping the Universe on the Pathway to SKA).

References

[1] Venter M and Bolli P 2017 *Int. Radio and Antenna Days of the Indian Ocean Conf.* (Cape Town)
[2] Mittra R and Galindo-Israel V 1980 *IEEE Antennas and Propagation Society Newsletter*
[3] GRASP, TICRA, Copenhagen, Denmark, [online] Available: www.ticra.com
[4] Spar Aerospace Limited 1981 *Feed Network 6/4-800-01 and -02 and Corrugated Horn CH-65C4* (Quebec)

IOP Conf. Series: Materials Science and Engineering **321** (2018) 012004 doi:10.1088/1757-899X/321/1/012004

Assessment of the Dominant Path Model and Field Measurements for NLOS DTV Signal Propagation

Geoflly L. Adonias and Joabson N. Carvalho

Department of Electrical Engineering
Federal Institute of Education, Science and Technology of Paraíba (IFPB)
Av. Primeiro de Maio, 720, Jaguaribe, João Pessoa - PB, 58015-435, Brazil

E-mail: gadonias7@gmail.com, joabson@ifpb.edu.br

Abstract. In Brazil, one of the most important telecommunications systems is broadcast television. Such relevance demands an extensive analysis to be performed chasing technical excellence in order to offer a better digital transmission to the user. Therefore, it is mandatory to evaluate the quality and strength of the Digital TV signal, through studies of coverage predictions models, allowing stations to be projected in a way that their respective signals are harmoniously distributed. The purpose of this study is to appraise measurements of digital television signal obtained in the field and to compare them with numerical results from the simulation of the Dominant Path Model. The outcomes indicate possible blocking zones and a low accumulated probability index above the reception threshold, as well as characterise the gain level of the receiving antenna, which would prevent signal blocking.

1. Introduction

The federal decrees no. 4,901 of 26th of November of 2003 and no. 5,820 of 29th of June of 2006 instituted and implemented, respectively, the Brazilian System of Digital Terrestrial Television [1]. At the first days of its implementation conducted by the Ministry of Science, Technology, Innovations and Communications and by the National Agency of Telecommunications, the focus was on the study and development of a high-performance system for the processing, transmission, and reception of digital audio and video signals. This structure should have the capacity to work side-by-side with the Analogue TV system while the transition period between both technologies [2, 3].

Those years of research were published as the Basic Plan of Distribution of Digital TV Channels, with the intention of defining parameters related to coverage and interference with current and future systems [2]. In 2007 the transmission of Digital TV signals started officially in Brazil, in the city of São Paulo, and since then, other cities over the country have been keeping a schedule defined by the Ministry of Communications for new deployments [1]. Because of the significant losses in the quality of services when contrasting digital and the analogue system, the reception power is an important aspect of the digital system [4].

Several prediction models were implemented as well as field measurements were performed around the cities [1, 2]. Those measurements are crucial in order to certify the prediction and to identify areas with a difficult reception. Computational modelling and field measurements are considered complementary approaches [1].

The study in [2], takes place in the main Brazilian capitals, making a comparison between field intensity measurements and numerical results from the implementation of ITU-R P.526, Deygout-Assis, ITU-R P.1546 and CRC Predict models. A similar study is conducted in [1] with 223 measurement points distributed along the metropolitan region of the city of Curitiba.

The purpose of this paper is to evaluate field measurements of Digital TV signal collected in the coastline of the city of João Pessoa and compare them with numerical results obtained from the implementation of the Dominant Path prediction model using the *WinProp Software Suite.*

2. Overview of the Dominant Path Model

Propagation models [5] in wireless communication systems have the objective of predicting the average received signal power at a given distance from the transmitter and, also, the inconsistency of the signal power in a close proximity to a certain location [6].

The Dominant Path Model (DPM) is based on the fact that not all power rays between the transmitter and the receiver contribute with similar energy to the total power received. In fact, only a few propagation paths are dominant in terms of energy input [7].

The DPM can be applied in indoor, urban and rural scenarios. It concentrates only on the dominant paths and does not calculate the paths with small energy contribution. Likewise, it does not consider all the details of the database, requiring less time to pre-process it [3].

The DPM model can be subdivided into two steps:

(i) Determination of the dominant paths (geometry);
(ii) Loss prediction along the paths.

The loss prediction (L) along the path is

$$L = 20\log\left(\frac{4\pi}{\lambda}\right) + 10\,p\log(l) + \sum_{i=1}^{n} f(\varphi, i) + \sum_{j=1}^{m} t_j - \Omega, \qquad (1)$$

where L is the path loss in dB, λ is the wavelength in meters, p is the visibility factor for *Line of Sight* and *Non-Line of Sight* conditions, l is the path length in meters, $f(\varphi, i)$ is a function which determines the interaction loss in dB, t_j is the transmission loss of wall number j and Ω is the wave-guiding factor [8].

In the case of mountainous urban scenarios, the topography of the area should be considered along with the height of the buildings for taking into consideration the visibility factor [8].

3. Materials and Methods

The applied methodology was based on the comparative analysis of field measurements and numerical results obtained from computer simulation using the Dominant Path Model.

The region of interest has an area of 7.73 km $\times$ 4.44 km, and it comprises all the neighbourhoods between the city centre and the coast line of *Cabo Branco* and *Tambaú.*

3.1. Field Measurements

A survey on the reception intensity of Digital TV signal in several points on the coastline of *Cabo Branco* beach and adjacent streets, in the city of João Pessoa, was performed.

Sixty-six measurements were collected *in loco* on the 9th of February of 2017 and divided into three routes: blue (first), green (second) and red (third) routes. The red route is 30 m higher than the other ones and the green route is right at the bottom of that 30-meter hill. Fig. 1 shows how those points were distributed along the routes.

TV Manaíra was the studied TV station, it serves as the João Pessoa city affiliate station of the *Rede Bandeirantes* and operates in the frequency range between 482 MHz and 488 MHz.

Figure 1. Measurement points (numbered from left to right).

3.2. Study of the Coverage Prediction Model

The coverage prediction was executed through the implementation of the Dominant Path Model using the *WinProp Software Suite.*

The simulation scenario was prepared by inserting the topographic information and by building the blocks that represent the various constructions in the region of interest so that the results could be as accurate as possible.

The antenna pattern and its azimuth angle were also taken into consideration as well as its transmission power and tower height. Specifications regarding the TV station can be found in details at the Brazilian National Telecommunications Agency system.

4. Results

The data collected in the field and through computational simulation were organised and assessed by each route. Figure 2 shows an overview of the whole area under simulation. The channel power, the buildings blocks and the location of the transmitter is indicated in both 2D (Figure 2(a)) and 3D (Figure 2(b)) representations.

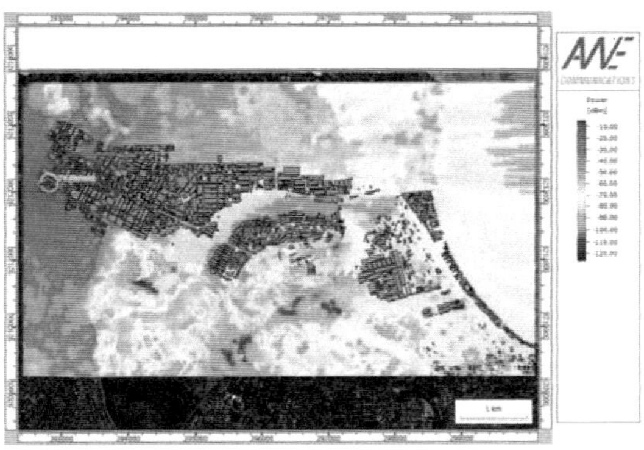

(a) Channel power in the region of interest.

(b) Scenario in 3D.

Figure 2. Channel power and 3D representation of the region of interest.

Figure 3 presents the data collected in the field and the data obtained from computer simulations for each of the three routes, Blue (Figure 3(a)), Green (Figure 3(b)) and Red (Figure 3(c)).

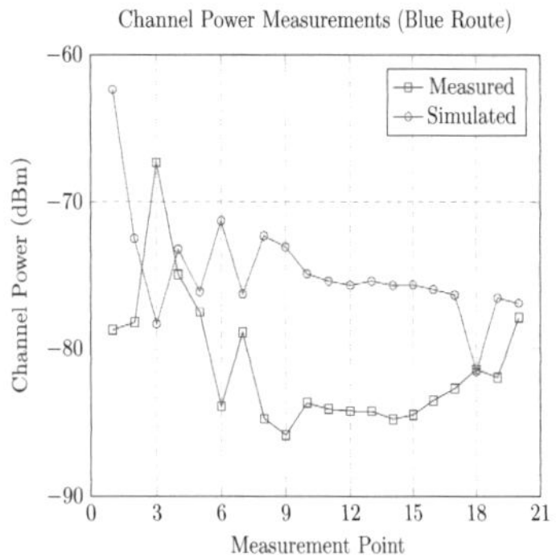

(a) Blue Route.

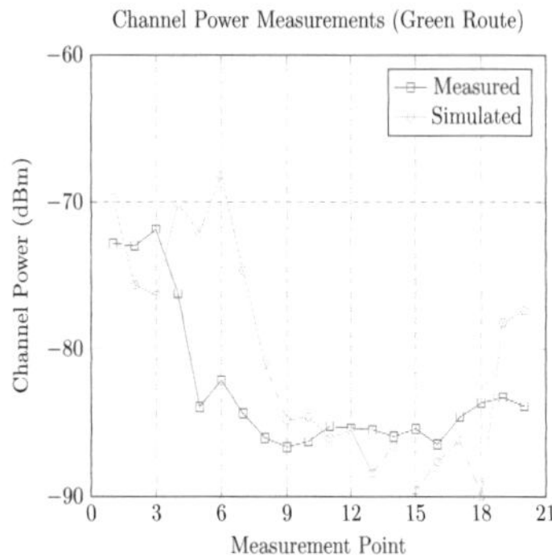

(b) Green Route.

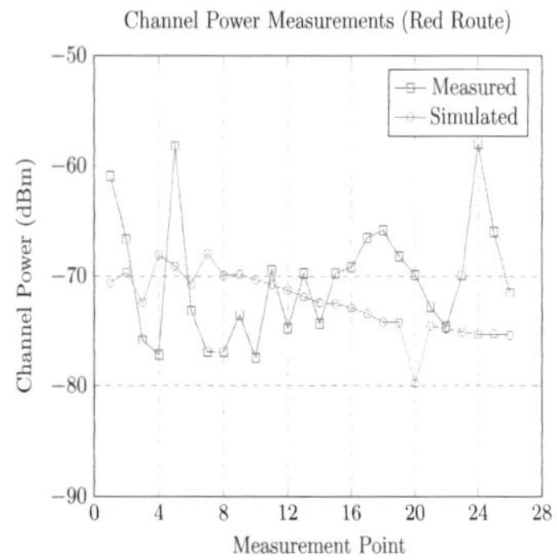

(c) Red Route.

Figure 3. Channel power measurements.

The numerical results, besides the comparative analysis, also calculated the accumulated probability index and it shows the probability in which a channel power might be below a certain reception power. On the condition that we are dealing with Digital TV signal, the reception threshold is −77 dBm, and it was calculated that the probability of it to be below this threshold is 20%.

5. Conclusion

In order to verify the accuracy of this work, line graphs were plotted (Figure 3). Figure 3(b) presents the worst scenario in which most of the measured and simulated points are below the reception threshold.

The accumulated probability index also shows us that only 80% of the region of interest has a coverage above the threshold of −77 dBm. Because the ideal percentage is to get as close as possible to 100% coverage, this is considered a very low percentage of the total region.

Based on these results, besides being a model that requires a low computational effort, the Dominant Path Model also showed itself as a reliable and accurate tool. For future works, it would be important to evaluate the same region of interest using other coverage prediction models.

References

[1] Vasco C L and de Borba M A C 2013 *Novas Técnicas e Ferramentas Aplicadas ao Sistema Brasileiro de TV Digital* ed Fonseca K V O and de Almeida Prado Pohl A (Curitiba, PR: Omnipax) chap 2, pp 39–88 1st ed ISBN 978-85-64619-14-2

[2] Silva F S, Matos L J, Peres F A C and Siqueira G L 2013 *Microwave Optoelectronics Conference (IMOC), 2013 SBMO/IEEE MTT-S International* pp 1–5

[3] Melo A A, Rodrigues F P, Farias K C S and Silva E F 2013 *XLI Congresso Brasileiro de Educação em Engenharia, 2013. COBENGE 2013. ABENGE*

[4] Arthur R, Iano Y, M S R and F R 2007 *IEEE Latin America Transactions* **5** 573–578 ISSN 1548-0992

[5] Sah N and Kumar A 2009 *International Journal of Computer and Electrical Engineering* **1** 464–472 ISSN 1793-8163

[6] Sharma H K, Sahu S and Sharma S 2011 *International Journal of Computer Applications* **19** 36–42

[7] Aschrafi A, Wertz P, Layh M, Landstorfer F M, Wolfle G and Wahl R 2006 *2006 IEEE 63rd Vehicular Technology Conference* vol 6 pp 2681–2685 ISSN 1550-2252

[8] Wahl R and Wolfle G 2006 *2006 First European Conference on Antennas and Propagation* pp 1–6 ISSN 2164-3342

Sea Turtles Geolocalization in the Indian Ocean: An Over Sea Radio Channel framework integrating a trilateration technique

Loic GUEGAN[1], Nour Mohammad MURAD[2] and Sylvain BONHOMMEAU[3]

[1]Univ Rennes, Inria, CNRS, IRISA, Rennes, France
[2]LE2P Laboratory, University of La Reunion, Saint Pierre, France
[3]IFREMER, Le Port, France

E-mail: `loic.guegan@inria.fr`

Abstract. This paper deals with the modeling of the over sea radio channel and aims to establish sea turtles localization off the coast of Reunion Island, and also on Europa Island in the Mozambique Channel. In order to model this radio channel, a framework measurement protocol is proposed. The over sea measured channel is integrated to the localization algorithm to estimate the turtle trajectory based on Power of Arrival (PoA) technique compared to GPS localization. Moreover, cross correlation tool is used to characterize the over sea propagation channel. First measurement of the radio channel on the Reunion Island coast combine to the POA algorithm show an error of 18 m for 45% of the approximated points.

1. Introduction

Assessing functional habitats for marine species is critical for conservation and management purposes. Since the 1990s, the development of tags that records the positions of marine species has enabled scientist to better determine these habitats. Sea turtles track are currently on the list of endangered species by the International Union for Conservation of Nature (IUCN). Hence, a better scientific knowledge about these species ecology is a critical need. French Research Institute for Exploitation of the Sea [1] wants to set up a localization system of sea turtles. There are different devices for the marine animal localization available on the market, but they are most often expensive and researchers have no access to the algorithms used to estimate animal positions. To overcome these disadvantages, we have chosen to use several radio modules, associated with different localization algorithms [2] that involve gateways to be placed on the coast. The distances between sea turtles and gateways will be in the order of kilometers, which is acceptable for this type of localization. Juvenile turtles that we focus the study on are generally foraging in coastal areas. Sea turtles can reach peak speeds of the order of 35 $km.h^{-1}$ and can have very short times spent at the surface to breathe between $[100\ ms; 500\ ms]$ and more longer times in $[2\ mins; 3\ mins]$ [3]. In addition, we take into account the radio conditions of the over sea environment, which may be extreme in some cases [4, 5] as waves, wind and sea spray. These elements alter the signal drastically. This is why we decide to model the over sea radio channel conditions, in order to adapt the Radio Frequency (RF) technologies and the algorithms used to attain the best possible accuracy on turtle localization.

Table 1: Channel profile of scenario 1 and scenario 2.

	Scenario 1		**Scenario 2**	
Paths	Delay (μs)	Mean gain (dB_m)	Delay (μs)	Mean gain (dB_m)
1	0	-0.38	0	-3.39
2	18.38	-5.00	11.73	-5.36
3	27.19	-4.73	26.89	-5.14

2. Benchmark and radio channel measurement

To model the over sea radio channel, we made a terrestrial measurement framework composed of a Turtle Transmitter (TT) and Gateway Receiver (GR). Software Define Radio technology is used to perform an easy and fast post-processing on the computer. Thus, the TT and GR devices are configured at a carrier frequency $f_c = 868\ MHz$ like the LoRa European Band with a transmit power $T_x = 9\ dBm$ with a sensitivity of $-148\ dB_m$. To realize a radio channel power profile we use a specific sequence called *Seq* of length $N = 456\ bits$ and processed with a sample rate $f_r = 10\ MHz$. Thanks to this, we are able to have a path resolution channel in the order of 100 ns. The *Seq* sequence is not encoded to protect itself and it's known by TT and GRs. Moreover, *Seq* is transmitted directly to consume the least amount of energy as possible and preserve the battery life on the TT. Also, for the same reason the transmit power T_x can not increased and is fixed. On the receiver side, a cross correlation algorithm is implemented along an observation window W to extract some channel profiles using a specific modulation like GFSK. Different experimental scenarios are done with the strong hypothesis that the TT is quite static for the moment and always on the ground. Also, multiple initial conditions are varied to analyze their influence on the RF signal propagation. First of all, the height h of the GR is varied in order to study its influence on the reception quality: multi-path, delay path, path loss. Then, we changed the distance d between the TT and the GR to highlight the signal attenuation.

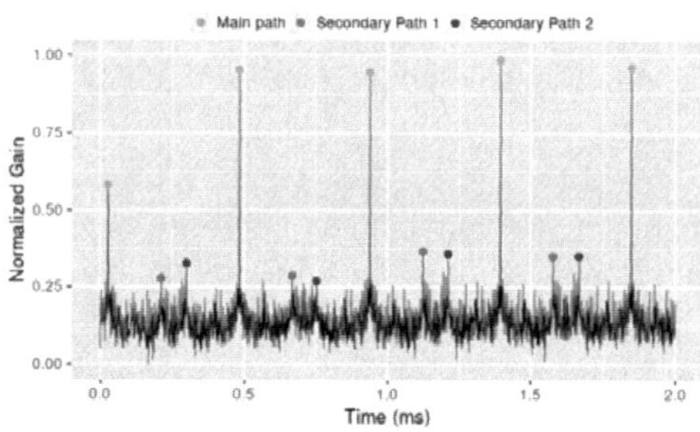

Figure 1: Scenario 1 normalized channel profile at $d = 20\ m$, $h = 1\ m$ and $W = 2\ ms$.

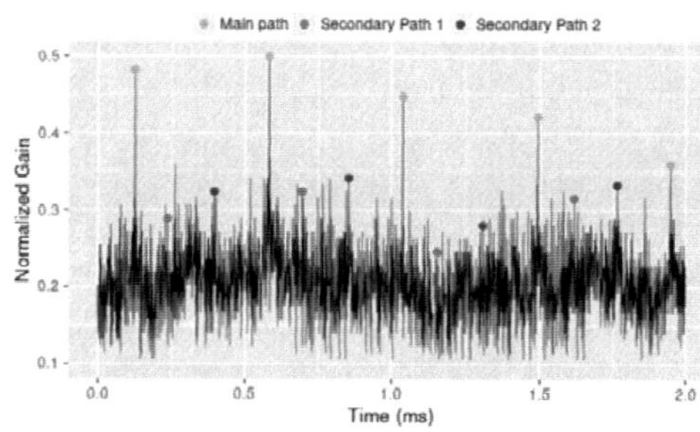

Figure 2: Scenario 2 normalized channel profile at $d = 20\ m$, $h = 0\ m$ and $W = 2\ ms$.

The results obtained by these two scenarios show the importance of the GR height. A more precise analysis of scenario 1 and 2 results shows that different secondary paths occurred and disturbed the main signal path. Table 1 summarizes the channel profiles for two scenarios.

Figure 3 presents the localization results realized on Omnet++ simulator combined with Matlab to simulate the propagation channel. On the left side of the figure 3 the localization is done by using an ideal channel called FreePathLoss and on the right side with the realistic channel model from the scenario 2. The algorithm used here is the Power Of Arrival (POA). The scenario 2 channel implementation gives more realistic results and this is why we integrate the over sea channel model. Actually, we did terrestrial experiments on radio beacon localization using POA algorithm. So, in a first time we introduce the POA algorithm used for the experiments then we present the first results for the terrestrial experiments.

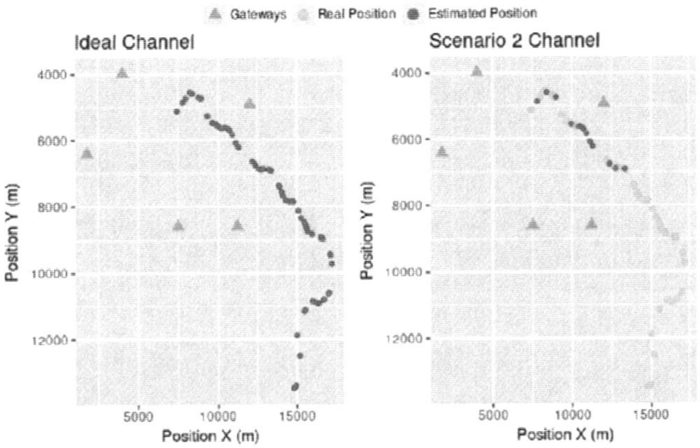

Figure 3: Localization of a turtle by trilateration using POA on ideal channel model and scenario 2 channel model.

3. Sea turtle localization algorithm based on PoA

To geolocalizate the sea turtles, we decide to use the multilateration [7] in order to approximate their relative positions to a given gateway called *master*. To be able to compute the distance between the sea turtles and the gateways, we do the strong hypothesis that we are in a line-of-sight situation in a classical free space path loss channel. Thanks to this hypothesis, we can approximate the distances between the sea turtles and the gateways. Thus, to do it we use a metric called the Receive Signal Strength (RSS). This metric corresponds to the receive signal power of each gateway. By knowing the signal power during the transmission and the RSS, it's possible to compute the distance between a given sea turtle and a given gateway using the following Friiz equation :

$$d = \frac{\lambda}{4\pi} \times \frac{1}{\sqrt{\frac{1}{P_t G_t G_r} \times 10^{\frac{P_r - 30}{10}}}} \tag{1}$$

With d is the distance between the transmitter and the receiver in meters; p_r is the Received Signal Strength in dB_m; p_t the Transmit power in Watt; G_t and G_r are respectively the linear gain of the transmitter antenna and the receiver antenna; λ is the signal wave length in meters. By knowing a minimum of three distances from a given turtle and the gateways, it is possible to estimate the sea turtle position by computing circle intersections. Each circle should be centered on a gateway with a radius that corresponding to the distance from this gateway to the sea turtle unknowing position. This is how the POA algorithm works. For validating the sea turtle approximated position, we decided to use the Global Position System (GPS) that provides the real sea turtle position at each transmission time with an error range between one and two meters. These values are based on real experiments. To be able to compare the position estimated by the POA algorithm and the GPS position, a projection of the geographic coordinates need to be done on a two dimensional plan. This projection should be relative to the master gateways for the comparison. The figure 4 presents the results obtained by a coastal experimentation. We use three gateways and one radio transmitter representing the sea turtle.

This figure shows also the paths followed by the turtle and its position at each transmission time. The purple points represent the differents approximated positions. So, the the algorithm is able to approximate 31 positions over 68. This represent 45% of the positions. For these approximated positions, we have mean error of 18 meters.

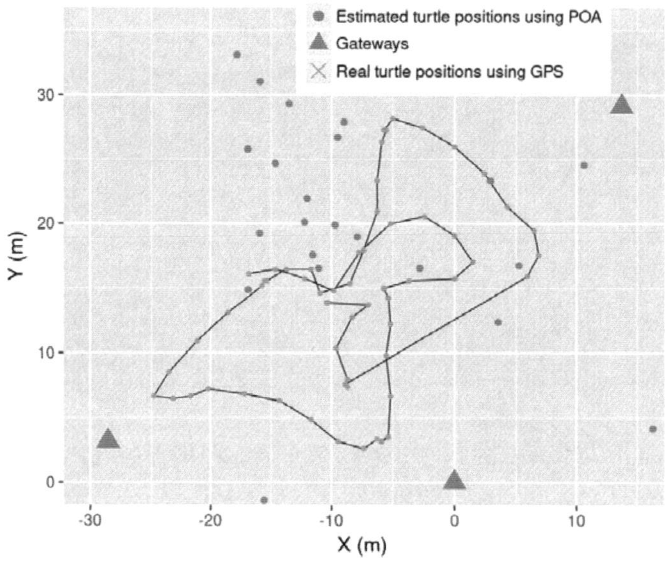

Figure 4: Results from POA experimentation.

4. Conclusion

In this article we show how to model an RF channel in real conditions using a correlation technique with a resolution of 100 ns. This approach has been validated for coastal purpose based on parameters such as the height or the distance. The height of the different gateways on the coast will be very important for the localization accuracy. Moreover our coastal experiment shows that localization in the over sea radio channel will not be easy. In fact, with line of sight communication channel we are able to approximate 48% of the total positions with 18 meters mean error. The results confirm that it is possible to geolocalize radio beacons using our technique for terrestrial purposes. The next step is to integrate the over sea radio channel to be able to do geolocalization experiments on sea turtles. To achieve this goal, we are planning to carry out measurement campaigns to be able to characterize the over sea radio channel. In this way, we will take into account the effects of the radio channel on the localization accuracy.

References

[1] "Ifremer", Ifremer.fr, 2017. [Online]. Available: http://www.ifremer.fr.
[2] F. Evennou, Techniques et technologies de localisation avances pour terminaux mobiles dans les environnements indoor, Universit Joseph-Fourier-Grenoble I, 2007.
[3] A. L. Southwood and al., Heart rates and diving behavior of leatherback sea turtles in the eastern Pacific Ocean, Journal of Experimental Biology, vol. 202, no. 9, pp. 1115-1125, 1999.
[4] F. Huang, X. Liao, and Y. Bai, Multipath Channel Model for Radio Propagation over Sea Surface, Wireless Personal Communications, vol. 90, no. 1, pp. 245-257, Sep. 2016.
[5] Y. Wu and al., Ray Tracing Based Wireless Channel Modeling over the Sea Surface near Diaoyu Islands, 2015, pp. 124-128.
[6] Semtech, Semtech SX1276 Datasheet.
[7] S. Ravindra et S. N. Jagadeesha, Time of Arrival Based Localization in Wireless Sensor Networks: A Linear Approach, Signal & Image Processing: An International Journal, vol. 4, n 4, p. 1330, aot 2013.

IOP Conf. Series: Materials Science and Engineering **321** (2018) 012006 doi:10.1088/1757-899X/321/1/012006

Distribution of rain height over subtropical region: Durban, South Africa for satellite communication systems

E O Olurotimi[1,5], O Sokoya[2], J S Ojo[3] and P A Owolawi[4]

[1,2]Department of Electronic Engineering, DUT. KZN. 4000, South Africa
[3]Department of Physics, FUTA, 340001, Nigeria
[4]Department of Computer Systems Engineering, TUT, Pretoria, 0001, South Africa

E-mail: elisayrot@gmail.com

Abstract. Rain height is one of the significant parameters for prediction of rain attenuation for Earth-space telecommunication links, especially those operating at frequencies above 10 GHz. This study examines Three-parameter Dagum distribution of the rain height over Durban, South Africa. 5-year data were used to study the monthly, seasonal, and annual variations using the parameters estimated by the maximum likelihood of the distribution. The performance estimation of the distribution was determined using the statistical goodness of fit. Three-parameter Dagum distribution shows an appropriate distribution for the modeling of rain height over Durban with the Root Mean Square Error of 0.26. Also, the shape and scale parameters for the distribution show a wide variation. The probability exceedance of time for 0.01% indicates the high probability of rain attenuation at higher frequencies.

1. Introduction

Statistical information of rain height is an important factor in rain attenuation prediction when designing satellite communication links, especially at higher frequency bands. Hydrometeors such as rainfall, hail, cloud, and rain height can frequently cause degradation of satellite signals, especially those operating at the frequency above 10 GHz. The result could be detrimental to the Quality of Service (QoS) at the receiving end [1-4]. Rain height is associated with the attenuation of the satellite signal and the interference from co-channel as a result of scattering at any location [5, 6]. In this work, an indirect procedure was employed due to the inadequate information of rain height in South Africa. Zero-degree isotherm height (ZDIH) obtained from the Tropical Rainfall Measuring Mission-Precipitation Radar (TRMM-PR) as provided in the product 2A23 algorithm have been used. ZDIH can be defined as the height where the temperature is at 0° C in the free atmosphere. ZDIH also determines the effective path length, which rain is affected. Though ITU-R Recommendation [7] provides the method of predicting rain attenuation which requires rain height, the recommendation also stated that around the world where there is no availability of specific information, the mean ZDIH should be considered as the proper quantity of ZDIH [8].

Several types of research have been carried out using numerous statistical distribution models over other atmospheric data such as rain rate, raindrop size distribution among others. However, very limited work has been done on ZDIH over South Africa. Hence, this work observes probability density function using one of the statistical distributions 'Three-parameter Dagum distribution' to model ZDIH based on the 5-year (2011–2015) data obtained from TRMM-PR over Durban, one of the largest cities in KwaZulu-Natal province of South Africa, 29° 97', 30° 95', altitude 8 m above sea level. Information on the classification of the season and data sources obtained from TRMM-PR is well presented in the work of [5, 9, 10]. The model was chosen among others because of its ability to define real-positive numbers over a wide range of empirical data [11]. Also, Root Mean Square Error was used to estimate the quality of statistical fitness.

2. Methods

2.1. Dagum Distribution

Dagum distribution was introduced in [11] for modeling data as an alternative to the Pareto and Log-normal models. This type of distribution has been extensively used for modeling in various data such as meteorological data, income, and financial analysis among others. Also, this distribution is among the continuous probability distributions that can be used to define real positive numbers over a wide range of empirical data [11]. In this work, three-parameter Dagum distribution is used and denoted as Dagum. The probability density function (PDF) and cumulative distribution function (CDF) of Dagum are presented as shown in equation (1) and (2) respectively:

$$f(x) = k\beta\alpha x^{-\alpha-1}(1 + \beta x^{-\alpha})^{k-1} \quad (1)$$

$$F(x) = (1 + \beta x^{-\alpha})^{-k} \quad (2)$$

where k and α are the shape parameters (α and $k > 0$), β is the scale parameter ($\beta > 0$) and x is the value of ZDIH ($x > 0$).

2.2. Dagum maximum likelihood method of estimation

The maximum likelihood method is one of the most regular methods for parameter estimation [12]. This method's success stems from its many valuable properties including reliability, asymptotic proficiency, invariance and simply its intuitive appeal [13]. Let $x_1, .. x_n$ be a ZDIH data size n from the Dagum distribution with parameters k, β, and $\propto$. The likelihood can be given as from equation (1):

$$L(k, \beta, \alpha) = (\beta k \alpha)^n \prod_{i=1}^{n} x^{-\alpha-1}(1 + \beta x^{-\alpha})^{-k-1} \quad (3)$$

and the log-likelihood function is also given as:

$$l(k, \beta, \alpha) = n\,(e - (b)) \textstyle\sum_{i=1}^{n} c \sum_{i=1}^{n} d \quad (4)$$

where $b = \alpha - 1$, $c = \ln x - (k + 1)$, $d = \ln(1 + \beta x^{-\alpha})$ and $e = \log k + \log \beta + \log \alpha$.

The maximum likelihood estimators of $\hat{k}_{MLE}, \hat{\beta}_{MLE}$, and $\hat{\propto}_{MLE}$ of the parameters k, β, and $\propto$ can then be obtained numerically by maximizing with respect to k, β, and $\propto$, in the log-likelihood function (4). In this case, the log-likelihood function is maximized by solving k, β, and $\propto$, x, in the non-linear equations which are given as follows:

$$\frac{\partial}{\partial k} l(k, \beta, \alpha) = \frac{n}{k} - \textstyle\sum_{i=1}^{n} \ln(1 + \beta x^{-\alpha}) = 0 \quad (5)$$

$$\frac{\partial}{\partial \beta} l(k, \beta, \alpha) = \frac{n}{\beta} - (k + 1) \textstyle\sum_{i=1}^{n} \frac{x^{-\alpha}}{(1+\beta x^{-\alpha})} = 0 \quad (6)$$

$$\frac{\partial}{\partial \alpha} l(k, \beta, \alpha) = \frac{n}{\propto} - \textstyle\sum_{i=1}^{n} \log x + (k + 1) \sum_{i=1}^{n} \frac{x^{-\alpha}}{(1+\beta x^{-\alpha})} = 0 \quad (7)$$

2.3. Performance Evaluation of the Distribution

Root Mean Square Error (RMSE) was used to estimate the quality of statistical fitness of distribution over the study location, as expressed in equation (8):

$$RMSE = \left(\frac{1}{n} \textstyle\sum_{i=1}^{n} \left(x_a(i) - x_p(i)\right)^2\right)^{1/2} \quad (8)$$

where n and i are the total numbers and cumulative rank of the data points, $f(x)$ is the density function of the fitted data, x_a and x_p are the measured value and Dagum predicted value of ZDIH respectively.

3. Results

3.1 Monthly variations of ZDIH

Figure 1 shows the variations of the monthly mean and corresponding 5-year monthly error bars of the standard deviation of the ZDIH. For example, Figure 1(a) shows that ZDIH increased as the year of data sampling increases, hence, the year 2011 and 2015 records the lowest and highest ZDIH respectively. The peak value of ZDIH observed in the year 2015 might be due to the presence of El Nino in the Southern part of Africa as it transits over the South Africa region. Also, Figure 1(b) revealed that the month of April presents the least ZDIH, followed by the months of February and March with the values of about 4.22, 4.31 and 4.32 km respectively. The standard deviation observed over these months is 0.05, 0.08, and 0.03. Consequently, the highest ZDIH transpired in the month of August followed by September and June with the value of about 4.46, 4.44 and 4.42 km respectively. The corresponding standard deviation is 0.08, 0.10 and 0.03 respectively.

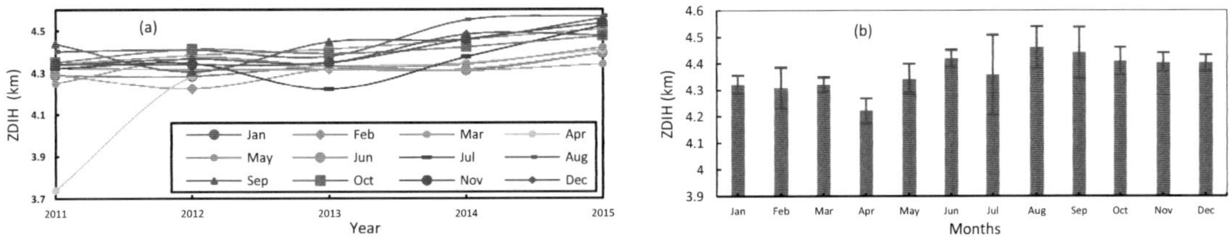

Figure 1. (a) Monthly mean, and (b) 5-year monthly mean with standard deviation for ZDIH.

3.2 Fitting of PDF and CDF to 0° C Isotherm height data

Figure 2 presents PDF and CDF fitting to the ZDIH over the average observation period based on equation (8) to estimate the quality of fitness. The results from Figure 2 show that Dagum present good fitting to the measured values of ZDIH with RMSE of about 0.26. Also, the parameter values estimated from MLE are 0.45, 131.65 and 4.42 for k, $\propto$, and β respectively.

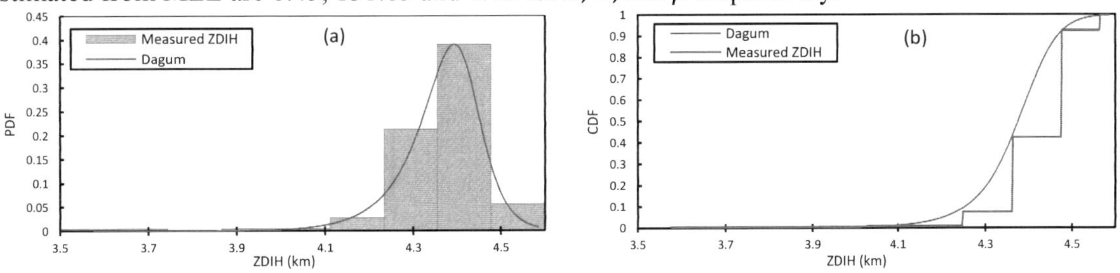

Figure 2. (a) PDF, and (2b) CDF of Dagum on the ZDIH data.

3.3 Mean Variations of the Dagum Distribution

Mean variations of monthly, seasonal, and annual Dagum distribution for the year of observation with the shape and scale parameters were presented in Table 1 and Figure 3. It could be seen from the Table 1 that both shape and scale parameters vary over a wide range of ZDIH values in the study location. For instance, in Table 1(a), the parameter k varies from 0.14 in April to 103.8 in August; α varies from 52.99 in the month of August to 135.2 in the month of October; while β varies from 4.02 in the month of August to 4.53 in the month of September. Also, Figure 3(a) has its lowest peak of 0.27 distribution in the month of December and the highest in the month of May with 0.56 peak distribution. Consequently, Figure 3(b) shows that the spring months have the highest distribution followed by winter; autumn and summer have the least distribution. Also, spring has the widest distribution followed by autumn, winter, while summer has the narrowest distribution. In Figure 3(c), the year 2014 has the highest and widest distribution followed by the year 2011 while the year 2013 has the lowest peak and narrowest distribution. This indicates the wide variations of ZDIH distribution based on the studied location.

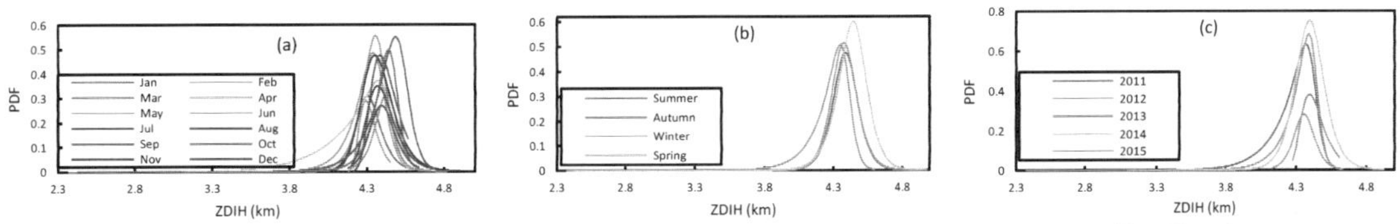

Figure 3. (a) Monthly Mean, (b) seasonal, and (c) annual variation of Dagum distribution on ZDIH.

Table 1. (a) Monthly, (b) Seasonal, and (c) Annual Dagum distribution of the parameters.

Months		Jan	Feb	Mar	Apr	May	Jun	Jul	Aug	Sep	Oct	Nov	Dec
	k	1.94	0.43	93.59	0.14	0.63	0.75	1.95	103.8	0.34	0.40	3.52	1.22
(a)	α	71.35	99.94	80.89	129.82	98.91	103.8	56.0	52.99	106.3	135.2	57.4	76.38
	β	4.26	4.38	4.06	4.44	4.38	4.44	4.30	4.02	4.53	4.47	4.28	4.38

Season		Summer	Autumn	Winter	Spring
	k	0.87	0.26	0.99	0.71
(b)	α	78.81	126.92	73.24	74.93
	β	4.40	4.40	4.38	4.47

Year		2011	2012	2013	2014	2015
	k	0.22	0.27	1.01	1.04	96.47
(c)	α	128	144.1	90.86	65.09	56.66
	β	4.42	4.43	4.36	4.39	4.055

3.4. Percentage exceedance of ZDIH

Figure 4 presents the percentage exceedance of the annual mean of ZDIH over the studied location. It could be noticed that ZDIH distribution varies from 3.58 to 4.59 km over probability range of all availabilities of signals. Also for 0.01% of time occurrence probability, about 4.56 km was recorded for ZDIH. This value corresponds to the high probability of rain attenuation at higher frequencies, especially at frequencies above 10 GHz.

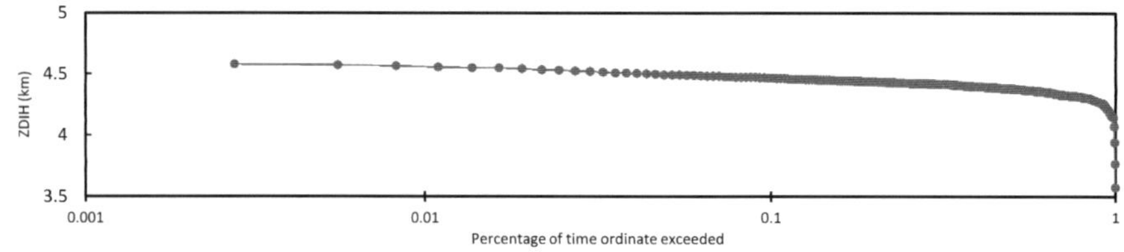

Figure 4. Exceedance probability of ZDIH.

4. Conclusion

This paper examined the statistical distribution of Dagum distribution to model zero-degree isotherm height over Durban in South Africa. Maximum likelihood method was used to derive the parameters of the distribution. Dagum distribution gives a good fit to ZDIH data with the RMSE of about 0.26. This technique had shown a good performance and may be useful for satellite system engineers, for link planning and systems design. The results further suggested the use of locally generated data rain height for the prediction of rain-induced attenuation.

5. References

[1] Dissanayake A, Allnutt J and Haidara F 1997 A prediction model that combines rain attenuation and other propagation impairments along Earth-satellite paths *IEEE Trans. on Ant. and Propag.* **45** pp 1546-1558
[2] Aris M, Azlan N, and Din J 2012 Rain height statistics from spaceborne radar for satellite communication in Malaysia *J. Teknologi (Sci. & Eng.)* **58** pp 1-5
[3] Ippolito L J 1986 *Radiowave propagation in satellite communications* (1st ed. New York: Van Nostrand Reinhold Company)
[4] Crane R K 1996 *Electromagnetic wave propagation through rain* (New York: Wiley-Interscience)
[5] Olurotimi E O, Sokoya O, Ojo J S and Owolawi P A 2015 Analysis of bright-band height data from TRMM-PR for satellite communication in Durban, South Africa *IEEE AFRICON 2015* pp 1-5
[6] Das S and Maitra A 2011 Some melting layer characteristics at two tropical locations in Indian region *XXXth URSI General Assembly and Scientific Symposium* pp 1-4
[7] ITU-R Rec. P.618-12 2015 Propagation data and prediction methods required for the design of Earth-space telecommunication systems *International Telecommunication Union, Geneva, Switzerland*
[8] ITU-R Rec. P.839-4 2013 Rain height model for prediction methods *International Telecommunication Union, Geneva, Switzerland*
[9] South Africa Weather. Available: http://www.sa-venues.com/no/weather.html
[10] Olurotimi E O, Sokoya O, Ojo J S and Owolawi P A 2017 Observation of bright-band height data from TRMM-PR for satellite communication in South Africa *J. of Atm. and Solar-Terrestrial Physics* **160** pp 24-33
[11] Dagum C 1975 A model of income distribution and the conditions of existence of moments of finite order *Proc. of the 40th session of the Int. Stat. Ins.* pp 196-202
[12] Casella G and Berger R L 1990 *Statistical inference* The Wadsworth & Brooks/Cole Statistics/Probability Series Wadsworth & Brooks ed: Cole Advanced Books & Software, Pacific Grove, CA
[13] Dey S, Al-Zahrani B and Basloom S 2017 Dagum Distribution: Properties and Different Methods of Estimation *Int. J. of Stat. and Probability* **6** p 74

IOP Conf. Series: Materials Science and Engineering **321** (2018) 012007 doi:10.1088/1757-899X/321/1/012007

Design and adaptation of a folded split ring resonator antenna for use in an animal-borne sensor

S. C. Dodson, P.G. Wiid, T.R. Niesler

Electrical & Electronic Engineering Department,University of Stellenbosch, Western Cape, South Africa

Email: sam.c.dodson@gmail.com, wiidg@sun.ac.za, trn@sun.ac.za

Abstract. We present the design, optimisation and practical evaluation of a folded split ring resonator (FSRR) antenna for the purpose of radio communication with an animal-borne sensor. We show that the measurements agree with the simulated results and that we are able to produce an electrically small antenna with low mismatch, high radiation efficiency and a quasi-isotropic radiation pattern. We then adapt the topology of the design from a circular to a rectangular shape, to completely fit inside the sensor enclosure. A quasi-isotropic pattern is maintained as well as low mismatch by appropriate tuning. There is a decrease in radiation efficiency which may be countered by a thinner substrate and retuning. We conclude that the adapted FSRR antenna is a suitable design for our application.

1. Introduction

The conservation and study of endangered wildlife can be assisted by real-time information obtained from animal-borne sensors. We consider the case of the African rhinoceros, where the sensor will be mounted on its lower back leg. For wireless communication with such a sensor, an antenna is required.

Since the sensor is a low-power device, the antenna needs to efficiently convert available power to radiated power. To realise this, the antenna should be well-matched to a 50 Ω line and internal ohmic losses should be minimised. The radiation pattern must be omnidirectional because the rhinoceros's orientation is variable. The antenna should be contained inside the sensor's enclosure, with dimensions of approximately $105\ mm \times 65\ mm \times 25\ mm$. These dimensions are much smaller than the $690\ mm$ wavelength at the operating frequency of $433\ MHz$, making the design challenging.

Little attention has been given to antenna design for tracking tags in the research literature. Usually, classic full-wave, half-wave and quarter-wave antennas are used. These radiators would not satisfy the required matching capabilities when scaled to the enclosure's dimensions. The folded split ring resonator (FSRR) antenna provides an attractive alternative for its ability to operate efficiently inside a small space, while satisfying matching, ohmic loss and pattern requirements [1]. We first consider the design of a prototype FSRR antenna based on the literature. Then we adapt the design to operate within the confines of the enclosure.

2. The folded split ring resonator antenna

The FSRR was first used as a radiator by Kim [1]. It is based on the single-arm split ring resonator which was conceived as a sub-wavelength metamaterial particle and has now seen extensive use as an electrically small antenna [2]. According to its definition, an electrically small antenna (ESA) fits inside its near-field [3]. The near-field radius for an ESA is given by the radianlength, $rl = \frac{\lambda}{2\pi}$, with λ the wavelength at the operating frequency [4]. This is also the inverse of the free-space wave number $k = \frac{2\pi}{\lambda}$. An ESA's size is typically given relative to the radianlength by the ratio $\frac{a}{rl} = ka$ with a the radius of the smallest fictitious sphere that totally encloses the antenna. ka is the antenna's electrical size.

The FSRR is depicted in figure 1. The radiation resistance R_{rad} (included in the total antenna input resistance R_{in}) and resonant frequency f_{res} decrease with smaller electrical size, hence an ESA will typically show high mismatch [5]. This is partly overcome by the folded conductor arms of the FSRR antenna which increase the input impedance Z_{in} by a factor of approximately four (compared to a single conductor) when the conductor widths are equal [6]. By raising the conductor width ratio $\frac{W_2}{W_1}$ above unity Z_{in} may be raised further. Additionally, the resonant frequency f_{res} may be lowered by decreasing the gap length GL. Both tuning mechanisms allow flexible antenna miniaturisation.

The FSRR antenna produces a quasi-isotropic radiation pattern. This behaviour may be interpreted by an even-odd mode current analysis [1]. The curved red arrows in figure 1 represent the positions of the maximum in-phase even mode currents, which form a y-directed electric dipole. The curved blue arrows represent the equal amplitude, in-phase odd-mode currents which form a z-directed magnetic dipole. The superposition of these two modes produces the quasi-isotropic pattern.

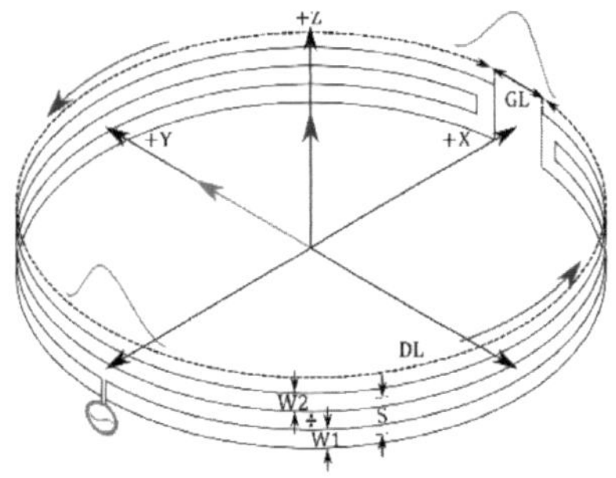

Figure 1. The FSRR antenna and its parameters.

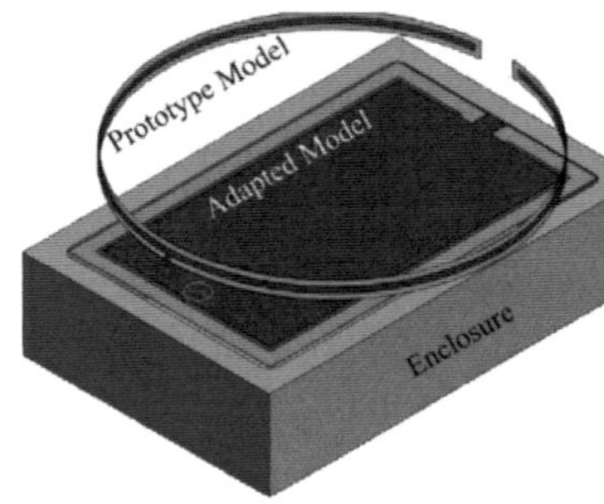

Figure 2. The first prototype and adapted prototype superimposed over the enclosure.

3. Optimisation and construction

A first prototype FSRR antenna was designed for performance validation. This was achieved using the computational electromagnetic software, FEKO. It was optimised to realise a 50 Ω input impedance on $0.2\ mm$ thick FR-4 substrate, with permittivity $\epsilon_r = 4.6$ and loss tangent $tan\delta = 0.017$. The gap length GL, dipole length DL, conductor width W (where both conductor widths were kept equal) and spacing S were optimised; their values may be seen to the left in table 1. The resulting electrical size is $ka = 0.48$, confirming that it is an ESA. Despite the small electrical size, the optimised prototype was too large to fit inside the enclosure. To adapt the model, it was flattened onto a rigid substrate and converted to a rectangular shape, as shown in figure 2. The antenna length TL is 305.87 mm as measured through the centre of the conductors. The shorter length, together with the shorter current path taken due to the sharp corners of the rectangle, causes f_{res} to increase from $435.5\ MHz$ to $471.4\ MHz$. R_{in} also decreases from $45.91\ \Omega$ to $32.63\ \Omega$. The electrical size of $ka = 0.55$ is larger than the first prototype's, but the adapted antenna's shape fits in the enclosure, regardless. To lower f_{res} the gap length was decreased from $10.4\ mm$ to $6\ mm$. For manufacturing reasons, the $0.2\ mm$ thick substrate was increased to $1.5\ mm$, which further lowered f_{res}. The conductor ratio W_2/W_1 was increased to 56.67 to raise R_{in} to $50\ \Omega$ at f_{res}. The parameters are shown to the right in table 1.

The first constructed prototype is shown in figure 3 (a). The flexible substrate made it difficult to construct accurately. It also required the construction of a large $\lambda/4$ sleeve balun [7]. The rigid substrate of the adapted prototype, shown in figure 3 (b), solved these issues. A small surface mount balun replaced the sleeve balun. The adapted prototype could also be machined to a much higher accuracy. The adapted feed was designed to face inward so that the sensor electronics could potentially be incorporated onto the substrate.

IOP Conf. Series: Materials Science and Engineering **321** (2018) 012007 doi:10.1088/1757-899X/321/1/012007

Table 1. Parameters of the first prototype and adapted prototype FSRR antennas.

First Prototype				***Adapted Prototype***				
DL (mm)	GL (mm)	W(mm)	S (mm)	TL (mm)	GL (mm)	W_1 (mm)	W_2 (mm)	S (mm)
315.8	10.40	1.38	3.53	305.87	6	0.72	40.8	2.66

(a) (b)

Figure 3. Constructed FSRR antennas: (a) the first prototype with the sleeve balun and metal plate highlighted in red and (b) the adapted prototype with the SMT balun highlighted in red.

4. Results

The aim of the first prototype was to determine whether the FSRR antenna could meet the project requirements. It would then serve as a benchmark for the adapted prototype. The adapted prototype should maintain the performance of the original, where possible. The measured and simulated S_{11} values of the first prototype are shown in figure 4 (a). The traces align closely, confirming the accuracy of the simulated model. The minimum S_{11} value of the measured model is $-17.44\ dB$ at $436.5\ MHz$. This is a good match to a $50\ \Omega$ line. At the same frequency the simulated radiation efficiency e_{rad} (which could not be measured for the first prototype) is 95.76% and hence internal losses are small.

The adapted prototype S_{11} traces are shown in figure 4 (b). The minimum value is $-19.77\ dB$ at $433.5\ MHz$, which is lower than the original prototype. This low value is expected, given that GL and $\frac{W_2}{W_1}$ were modified. It is suspected that the increased deviation between measured and simulated traces is because of the inner feed, which is not modelled accurately in FEKO. The difference between the minimum S_{11} values is $4.5\ MHz$, which can be accounted for in the design process. The measured e_{rad} is 61.04%, compared to the simulated 77.46%, which is much lower than the e_{rad} of the first prototype. It is caused by the increased substrate thickness. The e_{rad} may be increased by decreasing the thickness and reducing GL to compensate for the higher f_{res}.

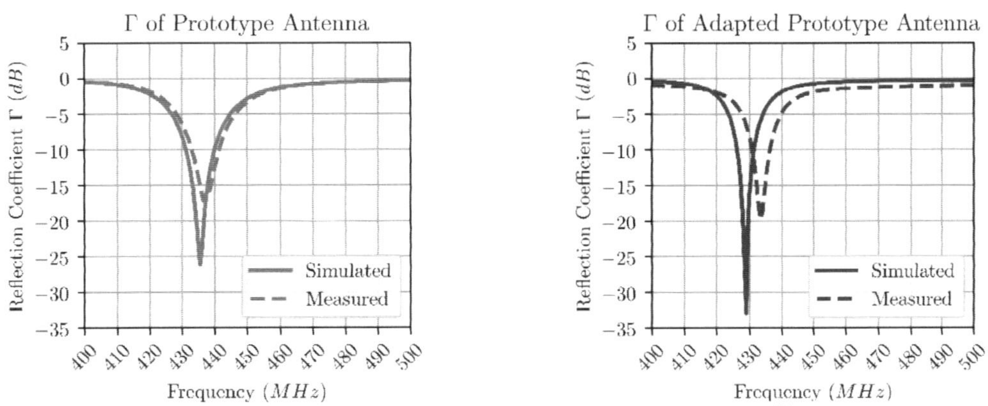

(a) (b)

Figure 4. Reflection coefficient plots of (a) the simulated first prototype FSRR versus the constructed model and (b) the simulated adapted FSRR prototype versus the constructed model.

Pattern measurements were conducted inside an anechoic chamber using a spherical near-field scanner. The antenna under test is obstructed for the positions $90° < \theta < 270°$. This corresponds to the lower portion of the polar plots which will be ignored. A metal plate was used to mount both antennas in the chamber, the plate is shown in figure 3 (a). The plate is sturdy and was designed for larger antennas. However, it leads to scattering and a plastic plate will be used for small antennas in the future.

The total directivity polar plots for the first configuration are shown in figure 5. The simulated model with no plate present is given by the solid trace. The pattern is quasi-isotropic with a maximum directivity of $1.26\ dBi$ toward $\theta = 90°$ in the XZ plane. The measured trace is dashed and the effect of the plate is noticeable; it reinforces the pattern toward $\theta = 60°$ and $300°$ while lowering it toward $\theta = 0°$. By including the plate in the simulation (dash-dotted trace), better agreement is obtained. We conclude that, without the plate, the antenna will radiate as predicted by the initial simulation.

The adapted prototype retained a similar main current path compared to the first prototype, hence it was speculated that the quasi-isotropic pattern would be maintained. This is confirmed by figure 5 (c) and figure 5 (d) which show the adapted model polar plots. The maximum directivity is in the same direction and is lowered slightly to $1.11\ dBi$.

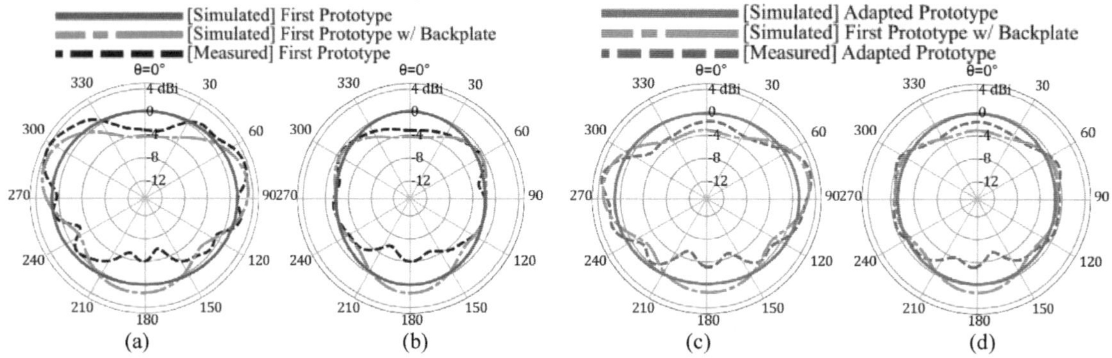

Figure 5. Far-field total directivity polar plots of measured and simulated antenna (at frequencies corresponding to their lowest S_{11} values), simulations with and without a 'plate', for (a) *XZ* plane first prototype; (b) YZ plane first prototype; (c) XZ plane adapted prototype; (d) YZ plane adapted prototype.

5. Conclusion & Future Work

An electrically small adapted prototype FSRR antenna of $ka = 0.55$ was designed, which satisfies most of the performance parameters and fits inside a sensor enclosure. It followed from an optimised first prototype, which satisfied the performance parameters, but was too large. The adapted model achieved a lower S_{11} value of $-19.77\ dB$ while retaining a quasi-isotropic radiation pattern. The adapted model is easier to manufacture and its inward-facing feed allows potential placement of sensor electronics on its substrate. The adapted prototype's measured e_{rad} of 61.04% is less than the simulated e_{rad} of the first prototype. However, this may easily be increased by using a thinner substrate and by decreasing the gap length. Future prototypes will analyse the effects of adding electronics to the substrate. A thinner dielectric will be used to increase radiation efficiency, and the antenna will be placed within the enclosure where its impedance characteristics will be measured.

References

[1] Kim J-H and Nam S 2017 A compact quasi-isotropic antenna based on folded split ring resonators *IEEE Antennas Wireless Propagation Lett.* **16** 294–97
[2] Dong Y and Itoh T 2012 Metamaterial-based antennas *Proc. IEEE* vol 100 pp 2271–85
[3] McLean J S 1996 A re-examination of the fundamental limits on the radiation q of electrically small antennas *IEEE Trans. Antennas Propagation* **44** 672–76
[4] Wheeler H A 1947 Fundamental limitations of small antennas *Proc. I.R.E.* vol 35 pp 1479–88
[5] Wheeler H A 1959 The radiansphere around a small antenna *Proc. IRE* vol 47 p 1325
[6] Balanis C A 2005 *Antenna theory: analaysis and design, 3nd ed.* (New Jersey: Wiley-Interscience) pp 517-15
[7] Saario S A, Lu J W and Thiel D V 2002 Full-wave analysis of choking characteristics of sleeve balun on coaxial cables *Electronics Letters* **38** pp 304-05

IOP Conf. Series: Materials Science and Engineering **321** (2018) 012008 doi:10.1088/1757-899X/321/1/012008

A Geometric Study of Tunable Planar Groove Gap Waveguide Cavities

Titus Oyedokun*, Riana Geschke*, Tinus Stander+

* Department of Electrical Engineering, Faculty of Engineering and the Built Environment University of Cape Town, South Africa

+ Carl and Emily Fuchs Institute for Microelectronics, Department of Electrical, Electronic and Computer Engineering,
University of Pretoria, South Africa

E-mail: riana.geschke@uct.ac.za

Abstract.
A comparative simulation-based study of tunable Planar Groove Gap Waveguide (PGGWG) cavities is presented. All cavities employ a simple biasing scheme to control the resonance of the cavity electronically. This is done by exploiting the DC isolation of the Planar Groove Gap Waveguide geometry for device biasing. Results obtained show that a higher tuning range (1.4%) is obtained with a rectangular PGGWG cavity as opposed to square and circular cavities. A higher unloaded Q-factor is obtained for the square cavity, which is more variable over the tuning range as compared to the circular and rectangular cavities.

1. Introduction

Planar integration media such as Substrate Integrated Waveguide (SIW) provide numerous advantages over hollow metallic waveguide, including low production cost and small size [1]. It is widely used in microwave systems where compact and moderately low-loss requirements are important [2]. Frequency-agile front-end filters are commonly implemented in SIW [3], [4]. However, due to the lack of DC isolation between the top and bottom conducting planes of SIW, bridging wires [5] or multi-layer routing [4] is required to connect the DC bias lines to the active device.

An alternative planar waveguide is the Planar Groove Gap Waveguide (PGGWG)[6]. PGGWG is an implementation of the Groove Gap Waveguide [7] in planar form. It is derived by replacing the electric sidewall with an artificial Perfect Magnetic Conductor (PMC) realised through an Electromagnetic Band Gap (EBG) structure [8]. This provides the advantage of DC isolation between the top and bottom plate, making the biasing of active device easier [9].

In this paper, we explore different tunable resonant cavity geometries in PGGWG using the biasing scheme as presented in [9] and compare them with respect to the achievable tuning range as well as the unloaded quality factor.

2. Planar Groove Gap Waveguide

The PGGWG structure is implemented on multilayer PCB, as shown in figure 1. The structure is formed by placing three rows of blind vias (diameter v_d), each capped by a catch pad at either

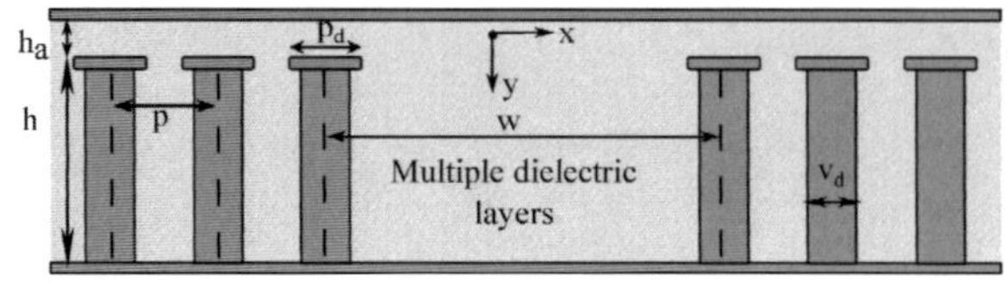

Figure 1. Cross-section of the Planar Groove Gap Waveguide. From [6].

side of a groove gap of width w. The blind vias and catch pads form an electromagnetic bandgap (EBG) structure.

The resonant frequency of the unit cell is determined by the capacitance between the catch pad and the bottom perfect electromagnetic conductor (PEC) plane, the capacitance between the catch pad and the top conductive plane, as well as the inductance of the blind via. By changing the size of the catch pad, the height and the width of the via, the bandgap that the EBG creates can be designed and shifted. This is demonstrated in figure 2 for the round top pad and square patch. Through appropriate choices of h and h_a, an artificial magnetic conductor (AMC) sidewall is created either side of the groove of width w. The medium then supports a TE_{10} propagating mode, similar to that of GGWG [10], in the EBG.

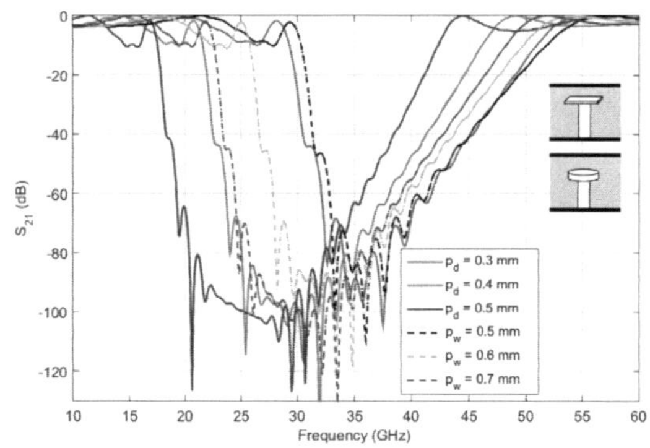

Figure 2. Magnitude of S_{21} for the different embedded EBG structures in parallel planes showing the effect of different pad and patch sizes. The insert is the EBG unit cell for round and square mushroom-type EBG where p_d is the diameter of the round pad and p_w the width of the square patch.

3. Geometry

3.1. Basic cavity

Various cavity arrangements, each formed with square and round EBG cells, of the PGGWG resonant cavity are explored. Figure 3(a), 3(b) and 3(c) show the rectangular, square and round resonant cavities with the corresponding electric field distribution of the fundamental resonant mode. Note the vias are topped by catch pads as detailed in figure 1.

The PGGWG cavity is excited using the CPW feed (Figure 3(d)) through weak input and output coupling in order to determine Q_0 of the resonant cavity while maintaining DC isolation between the two planes. The resonant frequency of the rectangular and square cavity is determined by the length of the cavity L_g while the coupling to the cavity, set by the iris width C_g, is intentionally chosen as a low value in this case, to minimize loading effects. The resonant frequency of the circular cavity is determined by its radius C_r.

Park Electrochemical Mercurywave 9350 low-cost substrate with ϵ_r of 3.5 is used in a three layer stack-up. The dimensions of the waveguide as presented in Figure 1 are h = 0.5 mm, h_a = 0.1 mm, v_d = 0.3 mm, pad diameter p_d = 0.7 mm (round pad EBG), p_w =0.7 mm and p = 0.85 mm. In the case of the rectangular cavity, w = 5.08 mm and L_g = 3.67 mm; for the square cavity, $L_g = w$ = 3.86 mm, and for the circular cavity the radius is C_r = 1.4 mm.

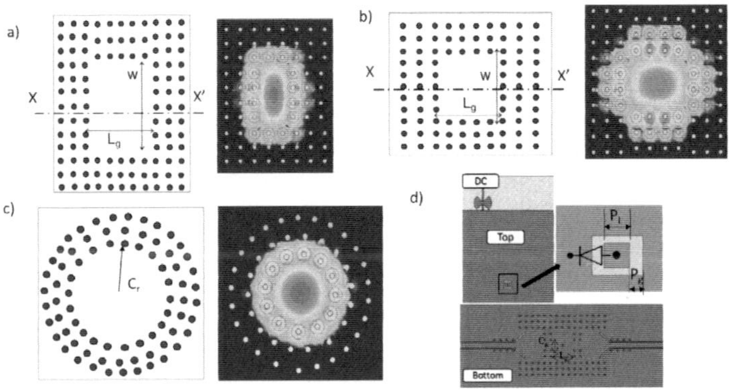

Figure 3. Electric field distribution for different resonant cavity arrangements. (a) Rectangular cavity. (b) Square cavity, (c) Circular cavity, (d) PGGWG structure with top and bottom view showing the connection of the varactor diode [9].

3.2. Varactor inclusion

A simple biasing scheme is applied as a result of the beneficial DC isolation between the top and bottom conducting plates of the PGGWG. As shown in figure 3(d), the DC bias line can be applied directly to the top metal plate of the PGGWG. This is done by etching a groove (P_g = 0.15 mm) around the square patch (P_L = 0.75 mm) in the center of the cavity. The patch is subsequently grounded to the lower conductor by a metallic via of diameter 0.3 mm, which creates an effective capacitive load on the resonant cavity at the local resonant E-field maximum. By varying the capacitance across the gap, the resonant frequency of the cavity is varied. To this end, a reverse biased varactor diode is connected across the gap, to electrically vary the gap capacitance. This approach is applied to all cavities.

4. Simulation result

To verify the achievable tuning range, a hybrid EM-circuit co-simulation is performed in CST Microwave Studio with the Spice model of a Macom 46600 series varactor diode. By varying the reverse bias voltage 0 - 30 V (corresponding to 0.13 - 0.25 pF) the results in Table 1 are obtained. Figure 4 shows the simulated S-parameters (insert shows the interface of the 3D model to DC bias line and varactor diode).

It is observed that the unloaded Q-factor is higher for square cavity than the rectangular cavity at the expense of lower achievable tuning range. On the other hand, the circular cavity has the least tuning range at 0.3% but with a fairly high and constant unloaded Q-factor.

5. Conclusion

A comparison between different electronically tunable resonant PGGWG cavity is presented using a simple biasing network. This simplified biasing scheme is a direct result of the PGGWGs unique property of DC isolated conducting planes. It is found that the unloaded Q-factor is fairly constant over the tuning range. Future work will investigate other cavity arrangements, and aim to increase the tuning range.

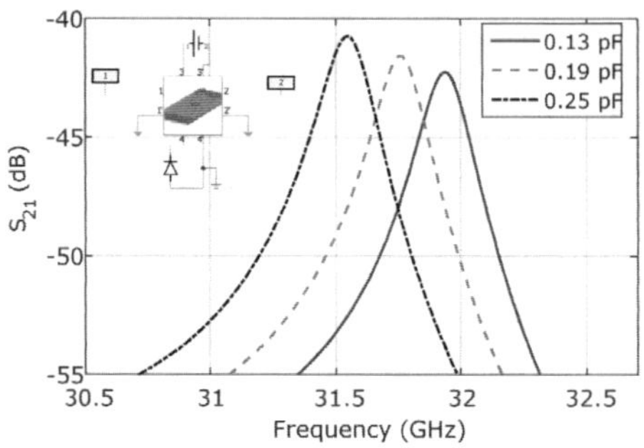

Figure 4. S-parameter of rectangular PGGWG cavity for different values of varactor capacitance

Table 1. Q-factor comparison between different resonant cavities

	f_0 (GHz)	Q_u	Tuning range %
Rectangular Cavity, Round EBG	31.54 - 31.95	152 - 160	1.4
Rectangular Cavity, Square EBG	38.48 - 38.81	136 - 137	1
Square Cavity, Round EBG	33.02 - 33.43	173 - 222	1.36
Circular Cavity, Round EBG	38.76 - 38.83	193 - 194	0.3

Acknowledgment

The use of CST Microwave Studio and NI AWR Microwave Office academic licenses are acknowledged. This work is based on research supported in part by the National Research Foundation of South Africa (Scholarship 99870 and Grant number 88100).

References

[1] Bozzi M, Georgiadis A, Wu K. Review of substrate-integrated waveguide circuits and antennas. IET Microwaves, Antennas Propagation. 2011 June;5(8):909–920.
[2] Tao Y, Hong W, Tang H. Design of A Ka-Band Bandpass Filter Based on High Order Mode SIW Resonator. In: 7th International Symposium on Antennas, Propagation EM Theory; 2006. p. 1–3.
[3] Entesari K, Saghati AP, Sekar V, Armendariz M. Tunable SIW Structures: Antennas, VCOs, and Filters. IEEE Microwave Magazine. 2015 Jun;16(5):34–54.
[4] Sirci S, Martnez JD, Taroncher M, Boria VE. Varactor-loaded continuously tunable SIW resonator for reconfigurable filter design. In: 41st European Microwave Conference; 2011. p. 436–439.
[5] He F, Chen XP, Wu K, Hong W. Electrically tunable substrate integrated waveguide reflective cavity resonator. In: Asia Pacific Microwave Conference; 2009. p. 119–122.
[6] Oyedokun T, Geschke R, Stander T. Experimental characterization of Planar Groove Gap Waveguide and Cavity. In: Proceedings of the 47th European Microwave Conference; 2017. p. 436 – 439.
[7] Rajo-Iglesias E, Kildal PS. Groove gap waveguide: A rectangular waveguide between contactless metal plates enabled by parallel-plate cut-off. In: Proceedings of the 4th European Conference on Antennas and Propagation; 2010. p. 1–4.
[8] Sievenpiper D, Zhang L, Broas RFJ, Alexopolous NG, Yablonovitch E. High-impedance electromagnetic surfaces with a forbidden frequency band. IEEE Transactions on Microwave Theory and Techniques. 1999 Nov;47(11):2059–2074.
[9] Oyedokun T, Geschke R, Stander T. A Ka Band Tunabale Planar Groove Gap Waveguide Cavity. In: IEEE Radio and Antenna Days of the Indian Ocean (RADIO); 2017. p. 1–4.
[10] Berenguer A, Fusco V, Zelenchuk DE, Snchez-Escuderos D, Baquero-Escudero M, Boria-Esbert VE. Propagation Characteristics of Groove Gap Waveguide Below and Above Cutoff. IEEE Transactions on Microwave Theory and Techniques. 2016 Jan;64(1):27–36.

IOP Conf. Series: Materials Science and Engineering **321** (2018) 012009 doi:10.1088/1757-899X/321/1/012009

FDSOI 28nm performances study for RF energy scavenging

E Rochefeuille[1,2], F Alicalapa[1], A Douyère[1] and T P Vuong[2]

[1]LE2P, EA 4079, Faculté des Sciences et Technologies, Université de la Réunion, 97744 Saint-Denis Cedex 9, Réunion

[2]IMEP-LAHC, UMR 5130, Institut Polytechnique de Grenoble, 38016 Grenoble Cedex, France

E-mail : edouard.rochefeuille@phelma.grenoble-inp.fr

Abstract. This paper presents a study on an integrated technology: Fully-Depleted-Silicon-On-Insulator (FDSOI) at a 28nm node. FDSOI results are compared to another technology: Complementary-Metal-Oxide-Semiconductor (CMOS) 350nm. The aim of this work was to demonstrate the advantages of using FDSOI technology in RF energy scavenging applications. Characteristics of transistors are pointed out and results showed an improved 22%-output voltage gain for a series rectifier and a 13%-output voltage gain for a Dickson charge pump in FDSOI technology compared to CMOS, for an input voltage and power of 0.5 V and 0 dBm respectively. Those results allowed to prove that FDSOI 28nm is a better technology choice for energy scavenging and low-power applications.

1. Introduction

Nowadays, microelectronic field aims to focus more and more on systems miniaturization as the reduction of power consumption without compromises on the required performances. Based on these objectives, some studies have been realized on energy conversion applications in [1], [2] and [3]. In addition to that, energy scavenging applications are considered to reduce the power consumption as in [4]. Moreover, reducing size could make evolve circuits like shown in [5]. RF energy scavenging is also present at nanoelectronic scale, which is designed to power integrated low-energy devices and sensors. The primary features and challenges of the FDSOI compared to CMOS has been described in literature as in [6]. Based on these facts, the aim of this study was to compare two technologies and to point out the advantage to use the emergent technology FDSOI in integrated applications such as RF-DC and DC-DC (charge pump) conversions applied to energy scavenging and low-power. This paper offers comparison simulations between CMOS 350nm and FDSOI 28nm technologies. Section 2 describes the remarkable transistors parameters. Then, the obtained output voltages of an RF rectifier and a Dickson charge pump structure [7] are discussed in section 3 and 4. Simulations were realized with Cadence Software by considering raw components. The ultimate goal of this study was to demonstrate that FDSOI 28nm is a suitable technology for RF energy scavenging.

2. Transistors characteristics

2.1. Transistors threshold voltages

This subsection presents studies of the drain current curves of different transistors of the technologies. FDSOI 28nm offers two interesting N-channel transistors: LVT (Low Voltage Threshold) and RVT (Regular Voltage Threshold). For CMOS technology, this work focuses on the

IOP Conf. Series: Materials Science and Engineering **321** (2018) 012009 doi:10.1088/1757-899X/321/1/012009

N-channel NMOS only. The following simulation gave the voltage threshold of the transistors by plotting the drain current Id for a drain voltage (Vds) of 1V and a gate voltage (Vgs) that varies from -1 V to +3 V. Results showed a voltage threshold of 200 mV for the LVT and 400 mV for the RVT in FDSOI 28nm. The same simulation was done for the NMOS transistor in CMOS 350nm. A threshold of 600 mV was found for the NMOS. These results show the best possible performances in FDSOI for low-power applications. Table 1 shows the threshold values.

	FDSOI LVT	FDSOI RVT	CMOS NMOS
Threshold Voltage	200 mV	400 mV	600 mV

Table 1. Threshold values of the studied transistors

2.2. Diode-connected transistor substrate connexion

This subsection presents a better way to connect the substrate connection of a four-connections transistor in diode-connected mode. The four-possible connexions are Gate, Drain, Source and Substrate. According to schematic on figure 2 and results on figure 1, the rectifying process performed better when substrate was connected to gate and drain. The advantages were therefore a good rectification with less current in negative part and a higher voltage. Figure 1 shows limits of the substrate connexion.

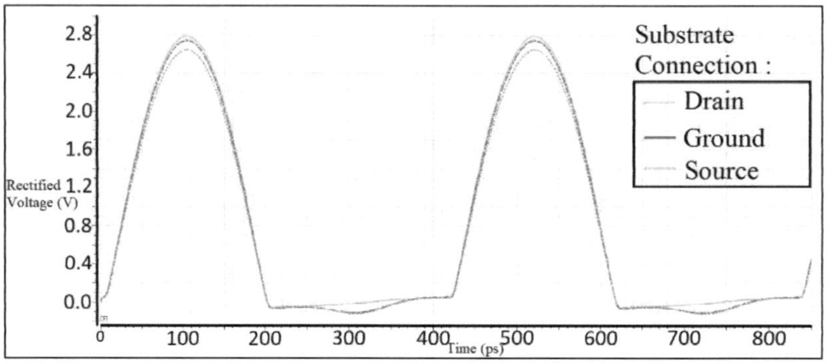

Figure 1. Transistors substrate connexion effect on rectified voltage

3. Rectifier

This section presents a comparison between simulated results obtained with a simple RF-DC rectifier, where the diode was replaced by LVT, RVT or NMOS transistors in diode-connected mode. Figure 2 shows the corresponding schematic.

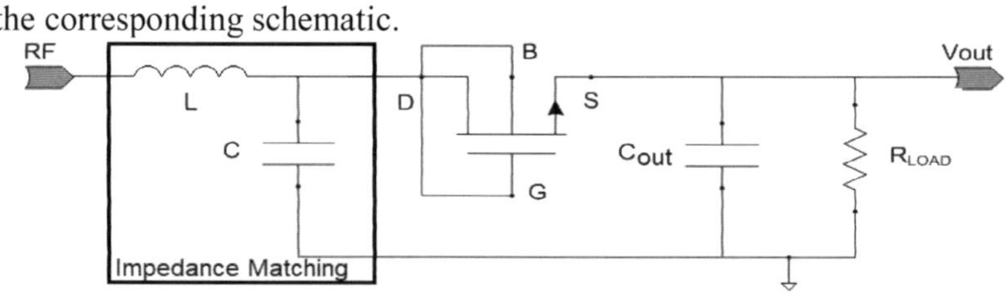

Figure 2. Schematic of a diode-series rectifier

The input power was set to 0 dBm at a frequency of 2.4 GHz. The output capacitance was set to 160 pF and the load resistance to 1 GΩ in order to show the best output voltages in this case. Figure 3a shows the output responses of the rectifier without using the impedance matching.

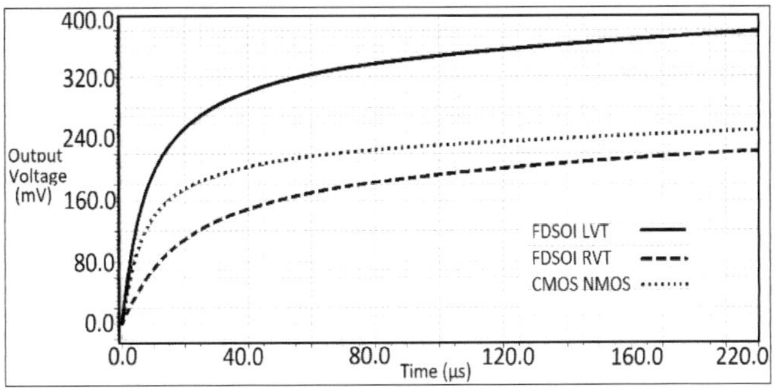

Figure 3a. Output voltages of rectifier

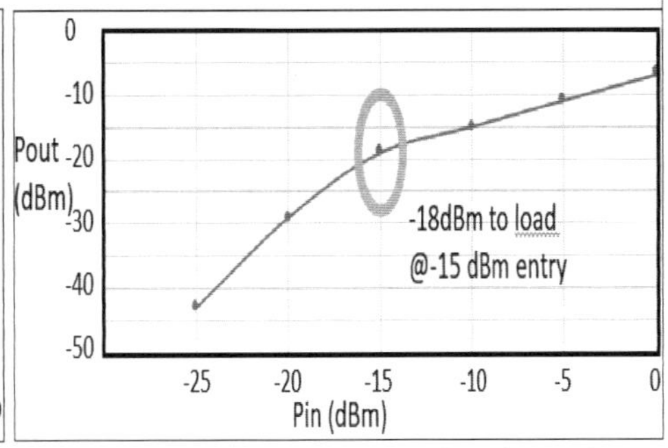

Figure 3b. Output power of an impedance matched rectifier

As shown on figure 3a, the use of a FDSOI LVT transistor allowed a simple rectifier to produce 390 mV output while the output voltage reaches only 240 mV in CMOS technology. The performance gain was therefore 22% on an RF simulated output voltage in favour of FDSOI technology. These performances came from the fact that FDSOI LVT threshold is lower than CMOS and this last parameter is linked to the rectifying diode-mode connection. The RVT transistor is not better than the NMOS in this case because it's default channel length is 4 times bigger than LVT. Channel length is a critical parameter in energy scavenging, more than channel width. Figure 3b shows the output power of a matched L-C rectifier for an input power that goes from -25 dBm to 0 dBm. The matching network allowed a maximum power transfer between the input generator and the transistor. A decent efficiency of 52% at -15 dBm was reached.

4. Charge Pump

This section presents a comparison between simulated results obtained with a classic 4-stages Dickson charge pump shown in [7] where the diode-connected transistors are replaced by LVT, RVT or NMOS. The input voltage was set to 0.5 V, the clock Vpp was 1V with a frequency of 100 KHz and the output resistance was 100 MΩ. Figure 4a shows the charge pump schematic and figure 4b shows the output responses of the charge pump.

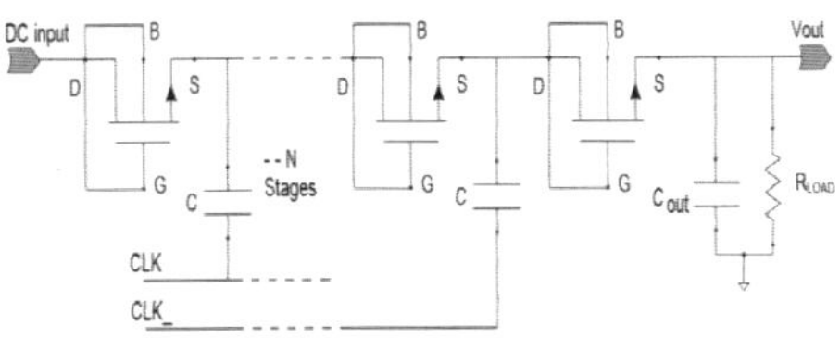

Figure 4a. Charge pump schematic

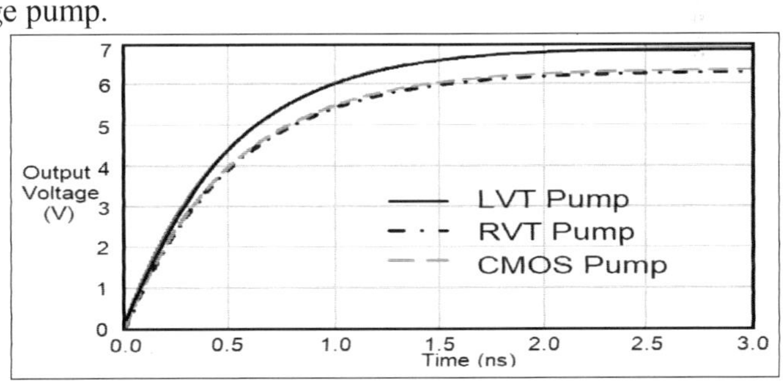

Figure 4b. Output voltages of charge pump using different transistors

As shown in figure 4b, the charge pump that operates with LVT diode-connected transistors achieved better performances than RVT and NMOS responses. The output voltage was 7 V for LVT charge pump while it was higher than 6 V for RVT and CMOS. In this case, a nearly 13%-output voltage improvement was scored by LVT response compared to CMOS because of its low voltage. In addition, it is important to mention that default dimensions of LVT are smaller than NMOS, typically a hundred nano-meters compared to a dozen of micro-meters.

To go further, a 4-stages Dickson circuit was simulated using optimal parametric values with FDSOI LVT. The pump performed up to 3V-output voltage from a 250mV-input with no load. An optimized circuit gave an output of 600 mV at 160 µA. Table 2 compares this work to literature.

Work	**[1] Moisiadis 2000**	**[2] Bhalerao 2007**	**[3] Chen 2010**	**This work**	
Type of work	Simulation	Simulation	1-Chip Integration	Simulation	
Process	Standard CMOS	Standard CMOS	65nm CMOS	28nm FDSOI	
Number of stages	4	4	3	4-Dickson based	
Input-Voltage	1.2 V	1.5 V	0.18 V	0.25	1.5 V
Output Voltage	3.5 V	6 V	0.5 V	0.6	11.4 V
Gain	2.9	4	2.78	2.4	9.5
Load	~58 kΩ	~133 kΩ	50 kΩ	3750 Ω	133 kΩ
Output-Power	~210 μW	~270 μW	5 μW	96 μW	969 μW

Table 2. Work comparison

Conclusion

This paper presents a study on two integrated technologies for RF energy scavenging. FDSOI 28nm and CMOS 350nm performances was compared. The FDSOI LVT transistor was particularly pointed out for its best voltage threshold and characteristics that allow higher output voltages over CMOS in low-voltage RF-DC and DC-DC conversions while reducing the overall dimensions. FDSOI technology is more expensive than CMOS but offers many advantages as a smaller die size and better performances. These comparison results lead to the choice of FDSOI technology for the design of a complete integrated rectifying module for RF energy scavenging.

Acknowledgement

The Autor would like to thank the Reunion Island Regional Council and the European Union – European Regional Development Fund (FEDER) PO 2014-2020 for their financial support to this work.

References

[1] Y Moisiadis, I Bouras and A Arapoyanni, "A CMOS charge pump for low voltage operation," *2000 IEEE International Symposium on Circuits and Systems*. Geneva, vol.5, pp. 577-580, 2000.

[2] S A Bhalerao, A V Chaudhary and R M Patrikar, "A CMOS Low Voltage Charge Pump," *20th International Conference on VLSI Design*, Bangalore, pp. 941-946, 2007.

[3] Po-Hung Chen *et al.*, "0.18-V input charge pump with forward body biasing in startup circuit using 65nm CMOS," *IEEE Custom Integrated Circuits Conference 2010*, San Jose, CA, pp. 1-4, 2010.

[4] T Paing, E A Falkenstein, R Zane and Z Popovic, "Custom IC for Ultralow Power RF Energy Scavenging," in *IEEE Transactions on Power Electronics*, vol. 26, no. 6, pp. 1620-1626, June 2011.

[5] A Douyère, S Rivière, J Rivière, F Alicalapa, J-D Lan Sun Luk, "Conception et réalisation d'un convertisseur RF/DC dédié à la collecte de faibles niveaux de puissance," *Journées Nationales Microondes*, May 2013, Paris, France, 2013, HAL<01137832>.

[6] D H Triyoso *et al.*, "Extending HKMG scaling on CMOS with FDSOI: advantages and integration challenges," *2016 International Conference on IC Design and Technology (ICICDT)*, Ho Chi Minh, pp. 1-4 2016.

[7] J F Dickson, "On-chip high-voltage generation in MNOS integrated circuits using an improved voltage multiplier technique," in *IEEE Journal of Solid-State Circuits*, vol. 11, no. 3, pp. 374-378, Jun 1976.

IOP Conf. Series: Materials Science and Engineering **321** (2018) 012010 doi:10.1088/1757-899X/321/1/012010

A Novel L-Probe Proximity Fed Patch Antenna With Parasitic Patch and Its Utilization in Antenna Arrays

Libor Sláma[12], **Josef Dobeš**[1], **Tomáš Boštík**[2], **and František Vejražka**[1]

[1]Department of Radioelectronics, Faculty of Electrical Engineering, Czech Technical University in Prague
[2]Tesla, Inc., Prague, Czech Republic

E-mail: poliux@seznam.cz, dobes@fel.cvut.cz, bostik.tomas@tesla.cz, vejrazka@fel.cvut.cz

Abstract. An analysis of the L-probe fed patch antenna with an extraordinary parasitic patch is described. The element of the antenna is fed by the L-probe partially implemented in PCB. An excellent impedance matching is obtained (< -26 dB in the design frequency band 4.4–5 GHz). The radiation characteristics are also very good (gain > 10 dBi). For the numerical analyses, the Full Wave — CST Microwave Studio software was used in both frequency and time domains, and a very good agreement between the Time Domain Solver (TDS) and Frequency Domain Solver (FDS) was obtained. Real antenna samples have been created and measured as well as eight-element antenna arrays designed by the Dolph-Chebyshev method.

1. Introduction

The microstrip patch antenna [1, 2, 3, 4, 5, 6] is a very popular choice for many wireless communication devices because of its low profile, light weight, and low cost, and it is mostly used in the Wi-Fi technology. The realization of the patch antenna could be very simple, but this simplicity leads to a narrow band (5 % for $|S_{11}| = -10$ dB). This is certainly acceptable for the Wi-Fi technology, but unacceptable for the direct wireless connections. Many techniques with the increase of the frequency band have been published, but only a few of them are useful for the direct wireless connections. The problem with most of them is that the gain of the antenna is decreasing or the direction of the main lobe is deflecting with frequency too much. This paper describes a novel structure which has acceptable properties in these two characteristics.

2. Broadband Patch

There are two fundamental possibilities of how to design the broadband patch antenna. The (natural) first of them is to decrease the Q-factor. The second one is to use a parasitic motive, or an appropriate combination of decreasing the Q-factor and using the parasitic motive. Both are discussed in 2.1 and 2.2.

2.1. Decreasing the Q-Factor

The relation between the Q-factor and impedance bandwidth (BW) is expressed by $\mathrm{BW} \approx 1/Q$, where the Q-factor can be evaluated as Energy_Stored/Energy_Dissipated. Combining these formulae, a simple way can be found to get the broadband antenna, which consists in an increase of the losses. Unfortunately, this technique decreases the gain of the broadband patch antenna. Hence, another method is to be found.

A more effective technique of the broadband-patch antenna design is decreasing the stored energy in the patch. If we imagine the patch element like a capacitor, its capacity is dependent on the distance of the conductive planes, and with the distance increasing, the energy decreases. (Fat substrate.) Unfortunately, when the patch is fed by the microstrip transmission line, increasing the described distance can contribute

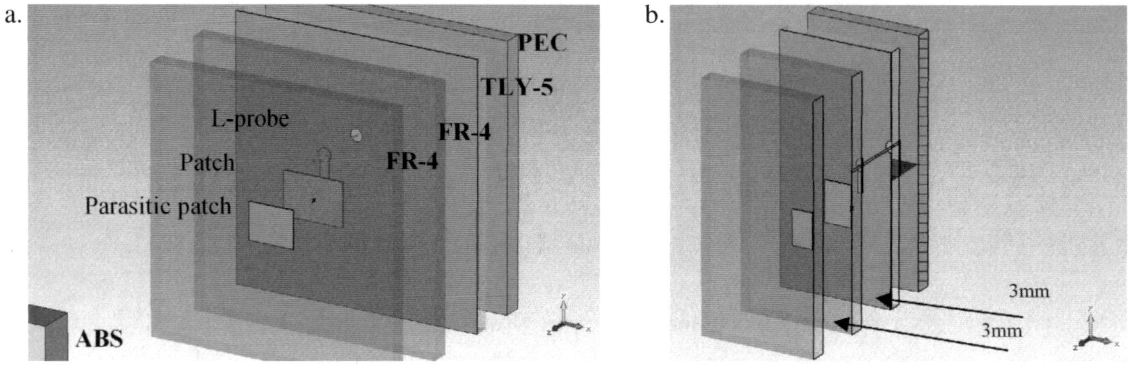

Figure 1. System with a parasitic patch and L-probe (a.) and cut through the structure with air gaps (b.).

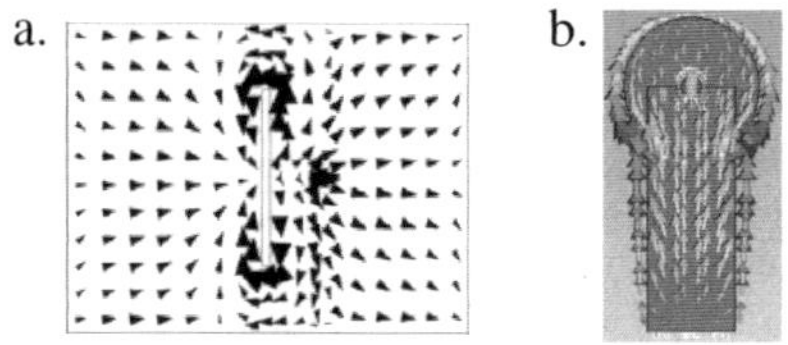

Figure 2. Current densities around the slot (a.) and current density on the L-probe (b.).

to excite the surface wave, and it is the reason why this parameter is a limiting factor. The first mode of the surface wave is the TE1 one with the cutoff frequency $f_m^{\mathrm{TE1}} = c/\left(4h\sqrt{\epsilon_r - 1}\right)$. Using the fat-substrate method, a construction of the microstrip feeding lines is impossible for a realization of low impedances.

2.2. *Parasitic Motives*

Inserting the parasitic motives to the patch antenna motive (slot) or inserting another patch near the first one (parasitic patch) can be exciting another resonance. In Fig. 2a, there is a two-dimensional graph of the current density on the patch around the slot. This current excites another antenna mode. However, this resonant need not radiates in a required direction. Therefore, finally, it can lead to a worse directivity.

3. Analysis of the Antenna Element and Array

First experiments have shown that using the microstrip transmission lines on the fat stack substrate leads to exciting the surface wave with the energy decreasing the radiation power.

In our novel proposal, the probe was located *under* the patch motive. Fortunately, this type of feeding the lines does not excite the surface wave as much as the microstrip line. Using this feeding line, a full separation of the radiating part and the feeding-line structures of the antenna has been achieved.

Finally, the L-probe [6, 2] feeding line has been used. For an easy manufacturability, the probe was particularly implemented in the bottom motive of the patch's PCB. The whole structure was completed by another PCB with the parasitic patch. The radome board was located in parallel to the PCB, and was

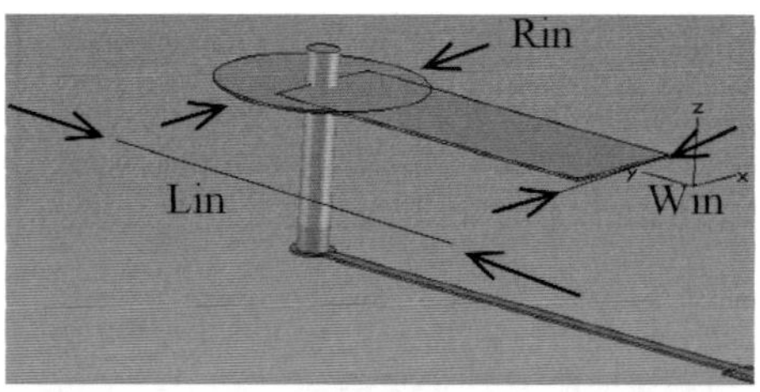

Figure 3. Detail of the L-probe implemented in substrate; Win = 4.42 mm, Lin = 11.92 mm, Rin = 6.7 mm.

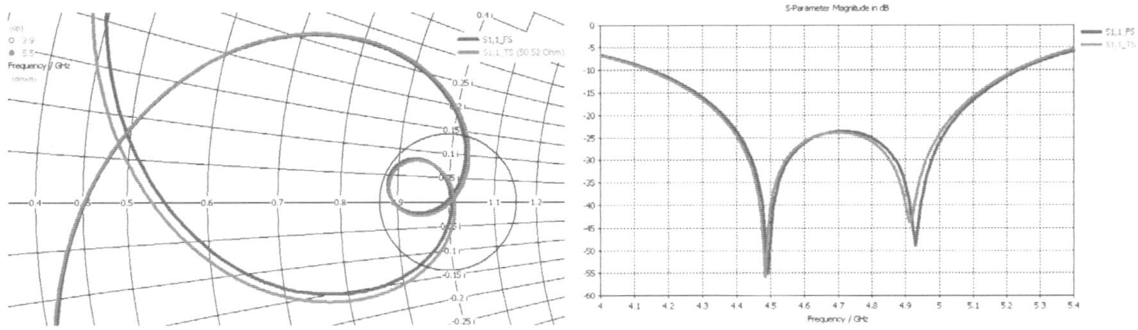

Figure 4. A TDS, FDS comparison for the structure with the transformer – Smith chart and magnitude.

simulated with a precise model. A picture defining the structure is shown in Fig. 1. The feeding line's impedance is 100 Ω. The thicknesses of the substrates are: FR4 – 1 mm, Taconic TLY-5 – 0.79 mm. The L-probe parameters in Fig. 3 were optimized to the best impedance matching. A metal ring motive connected to the pin was used to decrease the inductance of the probe. For Win = 4.42 mm, Lin = 11.92 mm, Rin = 6.7 mm, $|S_{11}| < -23$ dB has been achieved. A comparative simulation in FDS for this configuration has also been performed and a very good agreement of both simulations is shown in Fig. 4.

For the measurement of the sample, the structure was completed by the impedance transformer which transforms the antenna impedance to the standard one (50 Ω). Unfortunately, a bit narrower impedance band has been achieved – the results of these simulations are also presented here. However, the gain of the patch in the requested frequency band is still better than 10 dBi as clearly shown in Fig. 6.

For the true capabilities of the antenna, the simulations were also performed *without* the transformer. Mechanical parameters of both patches were optimized. The impedance dependencies in Smith chart can be seen in Fig. 5, where some dependencies of the impedance matching on h have also been explored.

The radiation patterns were also calculated for $\phi = 0°$ and $\phi = 90°$. In Fig. 7, some shift of the main lobe direction is seen for 90°. It is due to the current density on the feeding L-probe, which radiates as shown in Fig. 2b. However, the deviation is only 6.5°, which is acceptable. (There is *no* shift for $\phi = 0°$.)

Finally, an antenna array was created by the Dolph-Chebyshev method [3]. Eight-element arrays were made for the sidelobe levels −30, −25, and −20 dB with the parameters and radiation patterns in Fig. 8.

4. Conclusions

The proposed patch antenna exhibits a good performance in all the important parameters. The impedance matching is excellent and the radiating characteristics are also good. The eight-element antenna arrays yield gains around 20 dBi, 10 dBi more than the element. The antenna/arrays have been made by Tesla.

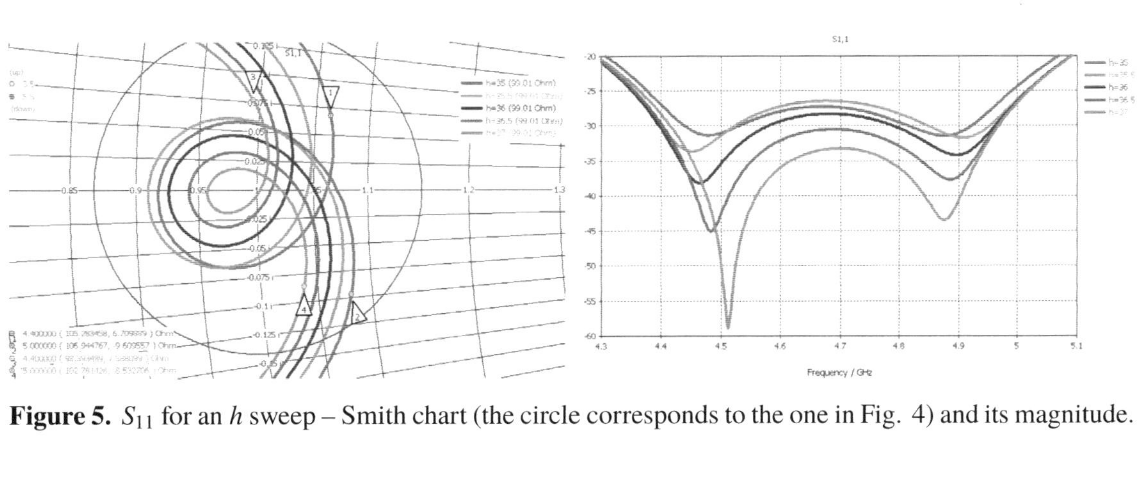

Figure 5. S_{11} for an h sweep – Smith chart (the circle corresponds to the one in Fig. 4) and its magnitude.

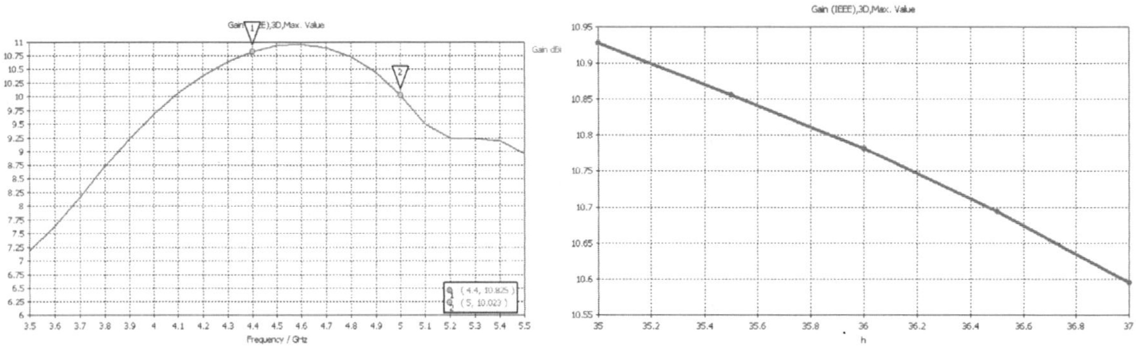

Figure 6. Gain of the patch antenna – frequency dependence and dependence on the height air gap (h).

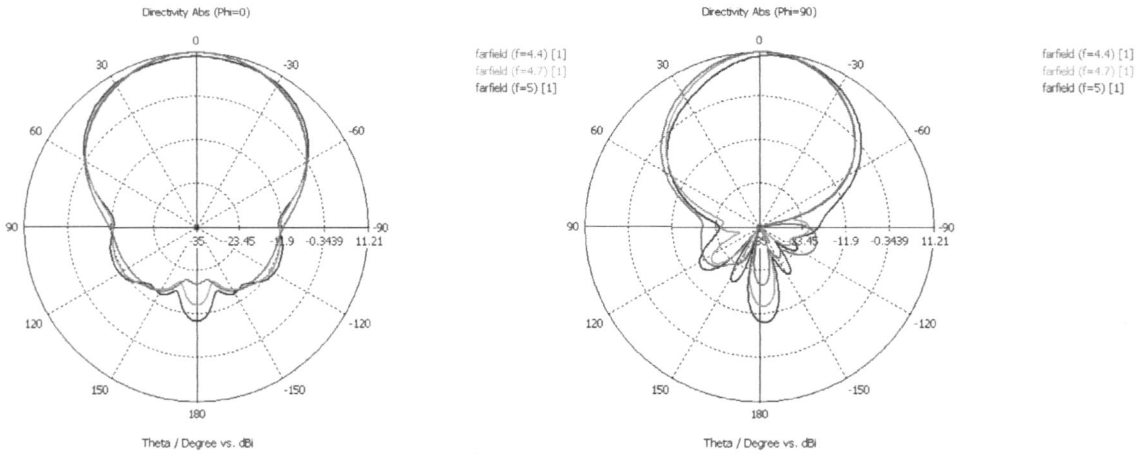

Figure 7. Radiation patterns for the cuts ϕ = 0° (left) and 90° (right) at f = 4.4, 4.7, and 5 GHz.

Element n.	Cheby -30dB	Cheby -25dB	Cheby -20dB
1	1.000	1.000	1.000
2	0.810	0.843	0.873
3	0.517	0.584	0.657
4	0.261	0.378	0.577

Design type	Cheby -30dB	Cheby -25dB	Cheby -20dB	Uniform
Gain (dBi)	19.2	19.5	19.8	20

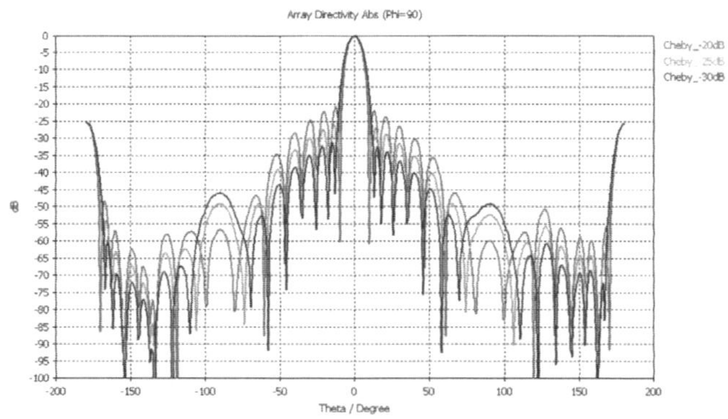

Figure 8. Antenna array – feeding weights, gains of arrays, and radiation patterns for used sidelobe levels.

Acknowledgment This paper was supported by the Technology Agency of the Czech Republic under the grant No. TE01020186.

References

[1] Balanis C A 2016 *Antenna Theory: Analysis and Design* 4th ed (New York: John Wiley & Sons, Inc.)
[2] Guo Y X, Mak C L, Luk K M and Lee K F 2001 *IEEE Transactions on Antennas and Propagation* **AP-49** 145–149
[3] Dolph L C 1946 *Proceedings of the IRE* **34** 335–348
[4] Zürcher J F and Gardiol F E 1995 *Broadband Patch Antennas* (Boston: Artech House, Inc.)
[5] Garg R, Bhartia P, Bahl I and Ittipiboon A 2001 *Microstrip Antenna Design Handbook* (Boston: Artech House, Inc.)
[6] Mak C L, Luk K M, Lee K F and Chow Y L 2000 *IEEE Transactions on Antennas and Propagation* **AP-48** 777–783

Institute of Physics
Dirac House, Temple Back
Bristol BS1 6BE UK

ISSN:1757-8981
ISBN 978-1-5108-6038-4